SpringerBriefs in Applied Sciences and Technology

SpringerBriefs present concise summaries of cutting-edge research and practical applications across a wide spectrum of fields. Featuring compact volumes of 50 to 125 pages, the series covers a range of content from professional to academic.

Typical publications can be:

- A timely report of state-of-the art methods
- An introduction to or a manual for the application of mathematical or computer techniques
- A bridge between new research results, as published in journal articles
- A snapshot of a hot or emerging topic
- An in-depth case study
- A presentation of core concepts that students must understand in order to make independent contributions

SpringerBriefs are characterized by fast, global electronic dissemination, standard publishing contracts, standardized manuscript preparation and formatting guidelines, and expedited production schedules.

On the one hand, **SpringerBriefs in Applied Sciences and Technology** are devoted to the publication of fundamentals and applications within the different classical engineering disciplines as well as in interdisciplinary fields that recently emerged between these areas. On the other hand, as the boundary separating fundamental research and applied technology is more and more dissolving, this series is particularly open to trans-disciplinary topics between fundamental science and engineering.

Indexed by EI-Compendex, SCOPUS and Springerlink.

Azman Ismail · Fatin Nur Zulkipli ·
Hairul Rizad Md Sapry · Mohd Ramzi Mohd Nor ·
Andreas Öchsner
Editors

Industrial Logistics I

Springer

Editors
Azman Ismail
Maritime Engineering Technology
Malaysian Institute of Marine Engineering Technology
Universiti Kuala Lumpur
Lumut, Perak, Malaysia

Hairul Rizad Md Sapry
Malaysian Institute of Industrial Technology
Universiti Kuala Lumpur
Johor Bahru, Johor, Malaysia

Andreas Öchsner
Faculty of Mechanical Engineering
Esslingen University of Applied Sciences
Esslingen, Baden-Württemberg, Germany

Fatin Nur Zulkipli
Faculty of Information Science
Universiti Teknologi MARA
Machang, Kelantan, Malaysia

Mohd Ramzi Mohd Nor
Malaysian Institute of Industrial Technology
Universiti Kuala Lumpur
Johor Bahru, Johor, Malaysia

ISSN 2191-530X ISSN 2191-5318 (electronic)
SpringerBriefs in Applied Sciences and Technology
ISBN 978-3-032-26980-5 ISBN 978-3-032-26981-2 (eBook)
https://doi.org/10.1007/978-3-032-26981-2

This Springer imprint is published by the registered company Springer Nature Switzerland AG
The registered company address is: Gewerbestrasse 11, 6330 Cham, Switzerland

Preface

The book delves into the multifaceted landscape of industrial logistics, offering insights into the critical issues and research trends shaping the field. With a focus on addressing challenges and proposing solutions, the chapters explore various dimensions of industrial logistics, including transportation, warehousing, inventory management, and supply chain integration. Drawing on a comprehensive review of existing literature and empirical studies, it highlights key issues faced by industrial logistics practitioners and researchers. Furthermore, the emerging research trends and innovative approaches were adopted in industrial logistics to overcome challenges and drive efficiency. These diverse strategies aimed at enhancing operational performance and sustainability in industrial logistics settings.

Lumut, Malaysia	Azman Ismail
Machang, Malaysia	Fatin Nur Zulkipli
Johor Bahru, Malaysia	Hairul Rizad Md Sapry
Johor Bahru, Malaysia	Mohd Ramzi Mohd Nor
Esslingen, Germany	Andreas Öchsner

Contents

Government Initiatives toward Consumer Intention to Purchase Hybrid Car Using the Extended Theory of Planned Behavior 1
Ahmad Fakhruddin Ahmad Dailami, Jimisiah Jaafar, and Sallaudin Hassan
1 Introduction 2
2 Literature Review 2
3 Methodology 3
4 Results and Discussion 4
4.1 Descriptive Analysis 4
4.2 Reliability Statistic 5
4.3 Correlation Analysis 5
4.4 Regression Analysis 5
5 Conclusion 6
References 6

Green Product Purchase Intention: Impact of Packaging Design, Printed Information, and Reusability 9
Ahzilah Wahid, Mohd Syahmi Mafizi, Nuur Fatin Roslan, Mohd Norazali Nordin, and Yong Salmah Nasir
1 Introduction 10
2 Research Objective 10
3 Literature Review 10
3.1 Green Purchase Intention 10
3.2 Packaging Design 11
3.3 Printed Information 11
3.4 Reusability 11
4 Conceptual Framework 12
5 Methodology 12
6 Conclusion 15
References 15

Investigation of Logistician Acceptance on Automated Vehicles in Logistics Industry 17
Amirul Haqim Khalid, Hairul Rizad Md Sapry, Helmi Adly Mohd Noor, Jamilahtun Md Ghazali, and Mohd Hilmi Abdul Rahman
1 Introduction 18
2 Methodology 19
3 Results and Discussion 19
4 Conclusion 21
References 22

Improving the Logistics Operation of Parcel Management through a Lean Management Approach 23
Hairul Rizad Md Sapry, Siti Atikah Zulkifli, Siti Norsharmimi S. M. Humayun Kabir, Mohammad Norizhan Nordin, and Abd Rahman Ahmad
1 Introduction 24
2 Methodology 25
3 Results and Discussion 25
4 Conclusion 28
References 29

Female Online Shopper's Trust: Examining Packaging Characteristics on Parcel Delivery Performance 31
Hairul Rizad Md Sapry, Nur Faqihah Rosli, and Abd Rahman Ahmad
1 Introduction 32
2 Methodology 32
3 Results and Discussion 32
4 Conclusion 35
References 36

Managing Lead Time of Shopping Good Handling Using Self-Service Kiosk 37
Hairul Rizad Md Sapry, Syafariena Rizzuwan, Siti Nurirdina Adilah Rahman, Sofia Alissa Mohd Saffei, and Abd Rahman Ahmad
1 Introduction 38
2 Methodology 39
3 Results and Discussion 39
4 Conclusion 42
References 43

Examining the Impact of Technology Adoption in Inventory Layout Design on Inventory Turnover ... 45
Hairul Rizad Md Sapry, Iza Nabila Sulaiman, and Abd Rahman Ahmad
1 Introduction ... 46
2 Methodology ... 46
3 Results and Discussion ... 49
4 Conclusion ... 49
References ... 51

The Role of Women in Leading a Technology Advancement in the Logistics Sector in Johor ... 53
Hairul Rizad Md Sapry, Aliah Suhailah Suhaimi, Mohd Yazid Md Taib, and Abd Rahman Ahmad
1 Introduction ... 54
2 Methodology ... 56
3 Results and Discussion ... 56
4 Conclusion ... 58
References ... 59

The Review and the Effect of Implementations: The Lean Six Sigma with Industrial Revolutions 4.0 in Logistics Industry ... 61
Hanafi Ahmad, Mohd Yazid Md Taib, Hairul Rizad Md Sapry, and Nurul Fadly Habidin
1 Introduction ... 62
2 Overview on IR 4.0 and Lean Six Sigma ... 63
2.1 Industry 4.0 ... 63
2.2 Lean Six Sigma ... 63
2.3 The Effectiveness of IR 4.0 and Lean Six Sigma on Supply Chains ... 64
3 Results and Discussion ... 64
4 Conclusion ... 65
References ... 65

A Bibliometric Analysis: The Effect of Lean Six Sigma 4.0 on Logistics Performance ... 69
Hanafi Ahmad, Ahmad Fakhruddin Ahmad Dailami, Mohd Yazid Md Taib, Hairul Rizad Md Sapry, and Nurul Fadly Habidin
1 Introduction ... 70
2 Overview on IR 4.0 and Lean Six Sigma ... 71
2.1 Industrial Revolutions ... 71
2.2 Lean Six Sigma ... 71
2.3 The Effectiveness IR 4.0 and Lean Six Sigma on Logistic ... 72
3 Methodology ... 72
4 Results and Discussion ... 73
4.1 Research Growth: Evolutions of Independency Between the IR 4.0 and Lean Six Sigma ... 73

4.2 Subject Area of IR 4.0 with the Lean Six Sigma ... 73
4.3 Geographical Distributions of IR 4.0 and Lean Six Sigma Publications ... 73
4.4 Bibliographic Network Analysis ... 73
4.5 The Impact Between LSS 4.0 and Logistics Performance ... 74
5 Conclusion ... 75
References ... 75

Factors Influencing the Demand for Commuter Service of Keretapi Tanah Melayu ... 77
Jamilahtun Md Ghazali, Haziq Alif Mohd Khairul Amin, and Hairul Rizad Sapry
1 Introduction ... 77
2 Methodology ... 78
3 Results and Discussion ... 79
3.1 Research Objective 1 ... 79
3.2 Research Objective 2 ... 81
4 Conclusion ... 83
References ... 83

Service Quality, Knowledge for Customer, and Customer Satisfaction Roles in Courier Services in Johor Bahru ... 85
Khairul Firdaus Adrutdin, Sallaudin Hassan, and Kamisan Gadar
1 Introduction ... 85
2 Background of Studies ... 86
3 Customer Satisfaction ... 88
4 Service Quality ... 88
5 Knowledge for Customer ... 89
6 Methodology ... 89
7 Results ... 89
8 Conclusion ... 92
References ... 92

Impact of Covid-19 Toward Consumer's Viewpoint in Food Delivery Services ... 93
Mazlan Md Zahid, Aminah Syakirah Ahmad Topik, Mohd Nasir Alias, Hairul Rizad Md Sapry, and Mustakim Melan
1 Introduction ... 94
2 Methodology ... 95
3 Results and Discussion ... 97
4 Conclusion ... 98
References ... 99

Using the E-servqual Model for Measuring E-Hailing and Taxi's Service Quality in Pasir Gudang, Johor. ... 101
Mazlan Md Zahid, Noor Shazwa Hajeerah Adnan, Mohd Nasir Alias, and Mustakim Melan
1 Introduction ... 102
1.1 Reliability ... 102
1.2 Responsiveness ... 102
1.3 Tangibility ... 103
1.4 Assurance ... 103
1.5 Empathy ... 103
2 Methodology ... 104
3 Results and Discussion ... 104
4 Conclusion ... 107
References ... 107

A Case Study: The Transportation Challenges in Agricultural Products at Lembaga Pertubuhan Peladang (LPP), Malaysia ... 109
Mohd Hilmi Abdul Rahman, Mohd Yazid Md Taib, Hairul Rizad Md Sapry, Nik Anis Zulaikha Mohd Shaharil, and Amirul Haqim Khalid
1 Introduction ... 110
2 Methodology ... 111
2.1 Research Procedure ... 112
2.2 Interview Approach ... 113
2.3 Data Analysis ... 113
2.4 Data Analysis Method ... 113
3 Results and Discussion ... 113
4 Conclusion ... 114
4.1 Reinstatement of Objectives ... 115
4.2 Recommendation ... 115
5 GPS Truck Tracking System ... 115
6 Improve Road Infrastructure ... 115
7 E-Commerce ... 116
References ... 116

Understanding the Islamic Hotel Concept from the Perspectives of Hotel Operators in Malaysia ... 117
Nor Amira Mohd Razali, Harnidah Samengon, Nur Hafizah Muhammad, and Nurashikin A. Ridzuan
1 Introduction ... 118
2 Methodology ... 119
3 Results and Discussion ... 120
4 Conclusion ... 122
References ... 123

Government Initiatives toward Consumer Intention to Purchase Hybrid Car Using the Extended Theory of Planned Behavior

Ahmad Fakhruddin Ahmad Dailami, Jimisiah Jaafar, and Sallaudin Hassan

Abstract The use of hybrid vehicles in Malaysia is reportedly accepted by consumers nationwide. The research about the Malaysian behavior toward hybrid car's is still insufficient. This study was conducted to find out the relationship between government initiatives, attitude in using the hybrid car, subjective norms toward the hybrid car, and the individual perceived behavior control with the consumer intention in Johor Bahru to purchase hybrid cars by focusing on the theory of planned behavior (TPB) extended framework. This study applied the quantitative research method where questionnaires have been distributed to 384 respondents in Johor Bahru. Correlation and regression analyses were conducted to examine the relationships between the identified factors and consumers' purchase intentions toward hybrid cars. The finding shows that government initiatives, attitude in using the hybrid car, subjective norms toward the hybrid car, and the individual perceived behavior control have a positive and significant relationship with intention to purchase hybrid cars in Johor Bahru, Malaysia. More research needs to be done to explore consumer behavior in promoting positive buying behavior toward the environment.

A. F. A. Dailami (✉)
Industrial Logistics, Malaysian Institute of Industrial Technology, Universiti Kuala Lumpur, Masai, Johor, Malaysia
e-mail: ahmad.fakhruddin@unikl.edu.my

J. Jaafar
Student Development, Malaysian Institute of Industrial Technology, Universiti Kuala Lumpur, Masai, Johor, Malaysia

Centre for Women Advancement and Leadership, Universiti Kuala Lumpur, Kuala Lumpur, Malaysia

J. Jaafar
e-mail: jimisiah@unikl.edu.my

S. Hassan
Malaysian Institute of Industrial Technology, Universiti Kuala Lumpur , Masai, Johor, Malaysia
e-mail: sallaudin@unikl.edu.my

A. Ismail et al. (eds.), *Industrial Logistics I*,
SpringerBriefs in Applied Sciences and Technology,
https://doi.org/10.1007/978-3-032-26981-2_1

Keywords Theory of planned behavior · Consumer intention · Government initiatives

1 Introduction

The presence of unwanted and dangerous substance into the environment can create the pollution [1]. The substance is harmful to the environment and human health. Some people do not neglect the responsibility of taking care of the environment. Several pollution resources are available in human surroundings that pollute the environment. Conventional combustion vehicles engine combusts the fuel and turn it into carbon emissions. Air pollution is harming the earth, humans, and wildlife. It can be considered as the most severe threat to the environment. It can lead to negative impact to human health by increasing the risk of heart disease, cancer, stroke, and other respiratory diseases [2].

With time and development, the automotive industry has changed to meet the demands of the contemporary world. Advancements have led to the design of automobiles that are long-term, cost-effective, safer, and more beneficial to the environment. Hybrid vehicles are one such technologies [3]. Hybrid cars have a lower environmental effect than internal combustion engine vehicles driven by petrol or diesel [4]. Malaysia is making a determined promise to reduce the intensity of its carbon emissions, namely a 40% reduction by 2020 and a 45% reduction in 2030 [5].

Green purchasing, also known as eco-friendly shopping, is the deliberate selection in purchasing the products or services that have a low environmental impact and encourage sustainability. It entails considering a product's whole life cycle, including its manufacture, usage, and disposal, to make educated decisions that are consistent with environmental ideals [6]. Companies are more encouraged to develop green products to meet the needs of this green advocation as a rising number of individuals recognize the necessity of individual responsibility for environmental conservation [7]. Therefore, it is crucial to research the variables influencing Malaysian consumers' inclination to acquire hybrid vehicles. Previous study has informed that future research about green products should focus on specific products like hybrid cars [8].

2 Literature Review

The topic of discussion centers on the elements that influence Malaysian consumers' intentions to buy hybrid vehicles, with particular attention paid to consumer behavior, attitudes, societal norms, perceived behavioral control, and government actions [9]. Due to their lower environmental effect, hybrid cars are becoming more and more popular among Malaysian consumers who are willing to pay a premium for environmentally friendly items [10]. Research indicates that beliefs, social norms, and perceived behavioral control all have an impact on consumers' intentions to purchase hybrid vehicles [11].

Due to individual and cultural variances influenced by experiences and values, opinions toward hybrid vehicles fluctuate [12]. Purchase decisions are highly influenced by favorable perceptions of green products and the caliber of eco-friendly goods [13]. Current owners of electric vehicles (EVs) are more likely to see EVs favorably in terms of cost-effectiveness and environmental benefits, according to studies including over 2000 participants [14]. Hypothesis 1: There is a positive significant relationship between attitude and consumer purchase intention toward the hybrid car.

A part of purchasing behavior is the reflection of subjective standards, which are societal pressures or expectations [15]. The social environments, cultural influences, and personal views of individuals impact the differences in these norms. Some consumers believe that driving a hybrid car is acceptable by society and that it is a wise decision to make in their social circles [16]. Hypothesis 2: There is a positive significant relationship between subjective norms and consumer purchase intention toward the hybrid car.

Consumer behavior is influenced by perceived behavioral control, which is characterized as one's confidence in carrying out an action in the face of difficulties [17]. People's intents to buy are influenced by their perception of their own ability to handle hybrid vehicles, which is based on their experiences and knowledge [18, 19]. Hypothesis 3: There is a positive significant relationship between perceived behavioral and consumer purchase intention toward the hybrid car.

Furthermore, government programs have a big influence on people's intents to purchase hybrid vehicles [20]. The government of Malaysia has implemented financial incentives, tax credits, and subsidies to lower the cost of buying hybrid cars [21]. Investments in EV charging infrastructure and technological advancements also support the adoption of environmentally friendly transportation [22]. Hypothesis 4: There is a positive significant relationship between government's initiatives and consumer purchase intention toward the hybrid car.

Overall, the study's hypotheses suggest positive relationships between attitudes, subjective norms, perceived behavioral control, government initiatives, and consumer purchase intentions toward hybrid cars in Malaysia. These factors collectively shape consumers' inclinations toward adopting environmentally friendly transportation options.

3 Methodology

For this investigation, a quantitative method was applied. A set of questionnaires has been adopted from previous research and distributed to 384 total respondents in Johor Bahru area at car dealers' showroom by using the non-probability sampling technique. By using this technique of sampling, the sampling process will become more convenient and easier to collect the data. The convenience sampling method was applied to gain the respondent. The questionnaire has been developed into three sections, A, B and C. Section A refers to the demographic questions such as age, gender, race, and more. Section B covers the respondent perspective for the independent variables. The last section covers the dependent variables question, which is the consumer intention in purchasing the hybrid car. Nominal scale was applied for the first section while section B and C have been designed by using the five points of the Likert scale. Cronbach's alpha has been run in determining data reliability. Multiple regression and correlation have been used to measure the relationship between the variables. The Statistical Package for the Social Sciences (SPSS) was implemented to process the data.

4 Results and Discussion

4.1 Descriptive Analysis

The research shows 384 respondents provided feedback for the survey. In the first profile, gender was mentioned. Women made up 67.2% of the study population, and men made up 32.8%. Within the study, the Malay made up 38.8% of all respondents, making them the ethnic group with the highest response rate. "Chinese" and "Indian" respondents make up 33.3% and 25.3% of the total respondents, respectively, by ethnicity. A further category called "Others" received 2.6% of the total respondents. The majority of those who participated in the research are the ages of 21 and 30 which representing 24.7% of all respondents. The sample's gain from respondents under the age of over 51 is the lowest (9.6%), while the sample's gains from respondents in the age brackets of under 20, 31 to 40, and 41 to 50, respectively, total 22.7%, 23.7%, and 19.3%. In addition, the sample from married respondents made up the bulk of the entire sample, accounting for 78.4% of it, while single respondents constituted about 21.6%. In terms of income, 57.6% of respondents had incomes of RM 4000 or more which represents the highest percentage of respondents' income levels. Secondly, this was followed by 19.5% of respondents with incomes of RM 3000 to 3999. In the meantime, around 16.4% of respondents with incomes less RM1999 and 21.4% of respondents had incomes between RM 2000 and RM 2999. The sample's educational background ranged from secondary school (SPM) to Doctorate holder. The highest sample academic qualification in this study, representing 48.4% of all respondents, was a bachelor's degree. The following are diploma, master's degrees, SPM, and PhDs, which represent 26.8%, 12%, 9.9%, and 2.9% of all respondents, respectively.

4.2 Reliability Statistic

The study found all the variables displayed Cronbach's alpha values greater than 0.7. All of the item's variables, it may be responded to, were reliable and consistent. So, it was appropriate for the investigation. There was a total of 16 items and five different variables. The perceived behavioral control has the highest Cronbach's alpha value, 0.920 (Excellent), with three items. Attitude toward purchasing a hybrid car with three items, 0.906, and consumer desire to buy a hybrid car with three items, 0.905, are great examples of Cronbach's alpha value. Good scores of Cronbach's alpha were found for the government's initiative (0.864) and subjective norms (0.885).

4.3 Correlation Analysis

All of the independent predictors were significant and have a positive linked to customer purchase intention for hybrid cars. The correlation analysis shows that all independent variables have high and positive correlation values. The study found the highest correlation value is subjective norms toward the consumer intention to purchase hybrid car which is 0.808. The lowest correlation value can be seen at consumer attitude toward intention to purchase hybrid car where the reading is 0.754. The readings for consumer's perceived behavioral control and government initiatives are 0.771 and 0.769.

4.4 Regression Analysis

The multiple regression analysis was carried out to understand the strength of the predictors toward the buyer's intention in purchasing the hybrid car in Johor Bahru. The study explains the model summary 0.862 is a strong value for effect size. There is also a strong correlation between the predictors with buyer's intention in purchasing the hybrid car in Johor Bahru. The study also has found that the predictors' variability is up to 74% of the dependent variable dimension. The finding was clearly significant as the ANOVA table's F-ratio illustrates how well the overall regression model fits the data. The research hypotheses are accepted as $p < 0.05$. From the data analysis, it concludes that there is an increase of 0.089 attitude toward purchasing the hybrid car (X1), 0.661 of subjective norms toward purchasing the hybrid car (X2), 0.257 of perceived behavioral control toward the purchasing the hybrid car (X3), and 0.144 of government's incentive toward hybrid car (X4) to increase 1 unit of consumer intention to purchase the hybrid car. The standardized coefficient explained the detail about the strength of the predictor. The result shows that the most effective influencer toward the consumer intention to purchase hybrid cars is the subjective norms (0.42) followed by perceived behavioral control (0.258), government initiative

(0.18), and attitude (0.088). Based on the SPSS result, the variables investigated in the current research were significant according to the hypothesis testing by correlation and multiple regression analysis.

5 Conclusion

In order to determine the predictors that influence the Malaysian consumer in purchasing the hybrid car, this research needed to be conducted. Previous research has analyzed consumer intention by using the theory of planned behavior [23]. The study examines the influence of consumer attitude, subjective norms, and perceived behavioral control toward the consumer intention in purchasing the hybrid car. Past studies have suggested that the future direction can be done by using the same theory and extend its dimension by studying another factors or variables. The government role toward influencing the consumer intention is important to study as it can bring a high impact toward the customer awareness [24].

Current research found that government initiatives, attitude in using the hybrid car, subjective norms toward the hybrid car, and the individual perceived behavior control have positive and significant relationship with consumer intention in purchasing the hybrid car in Johor Bahru area. This research was absolutely aligned with findings from previous research [25]. The study's empirical research demonstrates a positive relationship between government initiatives, attitude in using the hybrid car, subjective norms toward the hybrid car, the individual perceived behavior control, and a significant relationship with consumer intention in purchasing the hybrid car in Johor Bahru area. This finding is correlated with previous research conducted to analyze the role of government toward the Indian automobile industry [26].

The research has measured several limitations. Research has been conducted by using the convenience sampling method. As people are not randomly chosen from a larger population, sampling bias is possible. This might be impacting the findings of the research. Toward the theoretical implications, the study's results also imply that the extended TPB model is suitable for forecasting customers' intentions to acquire hybrid vehicles. The government is a key player in influencing the consumer intention to purchase the hybrid car. Future research can be conducted within the same area but exploring the new factors that impact the intention. Consumer decisions are influenced by other than the current studied variable. More research is needed in ensuring that greener products will be utilize in future.

References

1. National Geographic, Pollution. Education (2020)
2. WHO, *Ambient (Outdoor) Air Pollution* (WHO Global Health Observatory, 2022)

3. N. Asima, A.A. Bimani, A.S. AlSaqri, M. Matriano, The social and environmental impact of hybrid cars, in *Fourth Middle East College Student Research Conference*, Muscat, Sultanate of Oman (2019)
4. A. Porokhnya, A. Oganisyan, Hybrid cars—solution to ecological challenges; optimizing maintenance regimes for hybrid. IOP Conf. Ser. Earth Environ. Sci (2021)
5. KeTTHA, *The Green Technology Master Plan (GTMP)* (Ministry of Energy, Green Technology and Water, 2017)
6. P. Ansu-Mensah, Green product awareness effect on green purchase intentions of university students': an emerging market's perspective. Future Bus. J. **7** (2021)
7. L.F. Horani, L. Dong, Understanding sustainable purchase intention of smartphone users interface: evidence from Chinanvironmental psychology. Environ. Psychol. **14** (2023)
8. A.A. Mamun, Willingness to pay for environmentally friendly products among low-income households along coastal peninsular Malaysia. MDPI J. Sustain. (2018)
9. T.C. Wen, N.A. Noor, What affects Malaysian consumers' intention to purchase hybrid car? Asian Soc. Sci. **11**(26) (2015)
10. C.C. Patrice Seuwou, Measuring consumer behavioural intention to accept technology: towards autonomous vehicles technology acceptance model (AVTAM), in *Trends and Innovations in Information Systems and Technologies*, (2020)
11. V. Karuppiah, T. Ramayah, Modeling hybrid cars adoption using an extended version of the theory of planned behavior. Int. J. Transp. Res. **15**, 780 (2022)
12. I.A. Ajzen, Attitude-behavior relations: A theoretical analysis and review of empirical research. Psychol. Bull. **84**(5), 888 (1977)
13. F.Y. Millissa, W.T. Cheung, An extended model of value-attitude-behavior to explain Chinese consumers' green purchase behavior. J. Retail. Consum. Serv. (2019)
14. Gill, K.: Perceptions of electric vehicles are changing. Food Brooker News (2023)
15. B.T. Maqsood Hussain Bhutto, Predicting consumer purchase intention toward hybrid vehicles: testing the moderating role of price sensitivity. Eur. Bus. Rev. (2020)
16. N.S. Muhammad Iskandar Hamzah, Do pro-environmental factors lead to purchase intention of hybrid vehicles? The moderating effects of environmental knowledge. J. Clean. Prod. (2021)
17. C.A. Paul Sparks, The dimensional structure of the perceived behavioral control construct. J. Soc. Psychol. (1997)
18. M.N. Shahla Asadi, Factors impacting consumers' intention toward adoption of electric vehicles in Malaysia. J. Clean. Prod. (2021)
19. S. Kaffashi, M.N. Shamsudin, Transforming to a low carbon society; an extended theory of planned behaviour of Malaysian citizens. J. Cleaner Prod. (2021)
20. D.P. Ilias Geronikolos, An exploration of electric-car mobility in Greece: a stakeholders' perspective. Case Stud. Transp. Policy (2021)
21. Nor Ain Mohamed Radhi, H. N.: EVs and hybrids being considered as government department vehicles. The News Straits Times (2023)
22. U.A. Schmidt, Fuel Cell Electric Vehicles (FCEV): Policy Advances to Enhance Commercial Success. Sustain. Transp. (2021)
23. A. Vafaei-Zadeh, T.K. Wong, H. Hanifah, A.P. Teoh, K. Nawaser, Modelling electric vehicle purchase intention among generation Y consumers in Malaysia. Res. Transp. Bus. Manag. **43**, 100784 (2022)
24. M.I. Razak, A.M. Yusof, F. Mashahadi, Z. Alias, M.Z. Othman, Intention to purchase hybrid cars in Malaysia an overview. Int. J. Econ. Commer. Manag. (2014)
25. K. Shalender, N. Sharma, Using extended theory of planned behaviour (TPB) to predict adoption intention of electric vehicles in India. Environ. Dev. Sustain. **23** (2022)
26. S. Miglani, The Growth of the Indian Automobile Industry: Analysis of the Roles of Government Policy and Other Enabling Factors, in *Innovation, Economic Development, and Intellectual Property in India and China*, (2019)

Green Product Purchase Intention: Impact of Packaging Design, Printed Information, and Reusability

Ahzilah Wahid, Mohd Syahmi Mafizi, Nuur Fatin Roslan, Mohd Norazali Nordin, and Yong Salmah Nasir

Abstract This study aims to evaluate the effects of printed information, reusability, and packaging design on consumers' perceptions of and intentions to purchase green products. For this study, a questionnaire was used to collect data ($n = 170$). Respondents who lead green lifestyles and have purchased green products were selected at random using this technique. The data were analysed with SPSS. The results have practical implications for enhancing the performance of packaging for green products, based on the descriptive, correlational, and regression analyses from the SPSS results. The results demonstrate that intentions to make green purchases are positively impacted by printed information and reusability. Firms and businesses can also use this study to enhance their product marketing, boost consumer attitudes and knowledge about green products, and increase intentions to purchase green products. Good green product positioning is viewed as a competitive advantage that marketers can leverage to set their products apart from rivals in the market, creating the perception that their products are unique and thereby driving demand and purchase intention.

Keywords Reusable · Perception · Green packaged product · Printed information · Packaging design · Purchase intention

---

A. Wahid (✉) · M. S. Mafizi · N. F. Roslan · M. N. . Nordin · Y. S. Nasir
Maritime Management, Malaysian Institute of Marine Engineering Technology, Universiti Kuala Lumpur, 32200 Lumut, Perak, Malaysia
e-mail: ahzilah@unikl.edu.my

M. S. Mafizi
e-mail: syahmi.mafize@s.unikl.edu.my

N. F. Roslan
e-mail: nuurfathin@unikl.edu.my

M. N. . Nordin
e-mail: mnorazali@unikl.edu.my

Y. S. Nasir
e-mail: yongsalmah@unikl.edu.my

A. Ismail et al. (eds.), *Industrial Logistics I*,
SpringerBriefs in Applied Sciences and Technology,
https://doi.org/10.1007/978-3-032-26981-2_2

1 Introduction

Malaysian shoppers' intentions to purchase green products are increasing [1]. Green packaging, according to [2], is a growing issue amongst customers and merchants, and it is also associated with rising consumer knowledge of environmental sustainability. By concentrating on the three pillars of sustainable development, packaging design, printed information, and reusability become significant components that may promote advancements in the sustainable supply chain [3]. The very successful green packaging business is making great progress towards sustainability, accountability, safety, and low-cost management of green packaged products as consumer awareness of turning green and utilizing ecologically friendly products grows [4, 5].

2 Research Objective

This study aims to identify the factors that influence consumers' green product purchase intentions.

I. To identify the level of perception of packaging design, printed information, reusability, and green product purchase intention.
II. To identify packaging design, printed information, and reusability towards green product purchase intention.

3 Literature Review

3.1 Green Purchase Intention

Green product purchasing intent remains increasing amongst Malaysian purchasers' attitudes [1]. Agreeing to [2], green packaging is a growing issue amongst customers and merchants, and it is also associated with rising consumer knowledge of environmental sustainability. Through concentrating on the three pillars of sustainable development, packaging design, printed information, and reusability become the primary components that may promote advancements in the sustainable supply chain [3]. Through rising consumer awareness of turning green and adopting ecologically friendly products, the very successful green packaging business is making tremendous gains towards sustainability, accountability, safety, and low-cost management of green packaged products [4].

3.2 Packaging Design

In a competitive business market, packaging is crucial for attracting customers and influencing purchase decisions. Packaging is a crucial marketing tool that can significantly influence a consumer's purchase decision during the sale process [5]. It also serves as a conduit of communication between a product's image and the company's background [6]. In today's digitized and fast-paced world, consumers may easily search, compare, and pick their chosen items thanks to 'self-service' [4]. As a result, packing is significant, particularly from the standpoint of buying. Author of [7] further stated that packaging has many benefits, including the ability to draw in customers, convey important information about the product, categorize it in their minds, and finally set it apart from competing goods. Research from the past suggests that packaging can be customised in a variety of ways, including colour, design, graphics, and shape. These elements cause an effect on the consumers' choice of packaging.

Packaging design significantly influences consumer purchase intentions, especially for low-involvement products [2, 3]. It creates a brand image and stands out in display, as consumers rely on packaging design to make inferences about a product time constraint also influence purchasing behaviour [7, 8].

3.3 Printed Information

Printed materials provide product information and attract customers, with consumers focusing more on printed information when purchasing high-involvement products [9]. Consumers prefer to read printed information when purchasing health-related products, suggesting that consistency in the format of the information would make it easier for them to compare ingredients [10]. Conversely, research has indicated that there is no correlation between printed material and intentions to buy, particularly in developing nations [11]. In developing countries, consumers generally pay attention to the expiry date and ingredients of products. Several studies have found that printed information on products tends to stimulate consumer purchase intentions [12].

3.4 Reusability

Authors of [13] highlight of the benefits of packaging that can be reused several times and is regularly used for the transportation of commodities. It is often built for lifespan, simplicity of use, cleanliness, and reparability, and it is preferable foldable or nest able to maximize space efficiency [7]. All of these characteristics make it easy to return to the original user, who will be able to utilize it in a timely way. Reusable packaging is designed and built to last a long time. It should preferably be recyclable once it has served its purpose, preventing rubbish from ending up in a landfill [14].

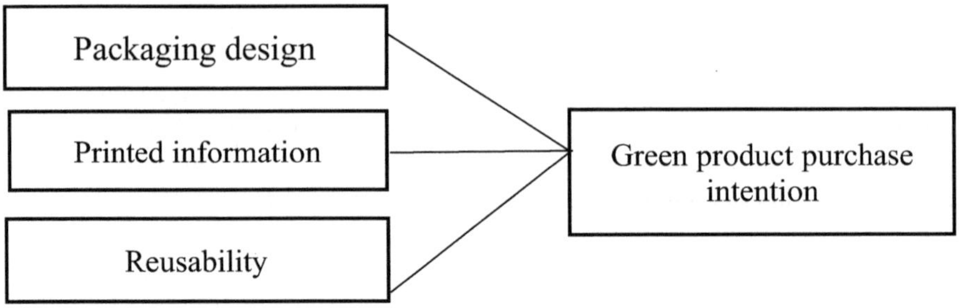

Fig. 1 Conceptual framework

4 Conceptual Framework

The conceptual framework is a research structure that supports and connects various aspects, adapted from [10]. The framework in Fig. 1 suggests that packaging design, printed information, and reusability are potential factors influencing green product purchase intention.

5 Methodology

The information was gathered using a Google form. Respondents were briefed on the study before being guided through the questionnaire using the Likert scale. To meet the study's objectives, each portion was to be analysed. These studies were based on samples of university students. Referring to the appendix, Table 1 outlines the demographic profile of respondents, encompassing gender, marital status, age groups, and education levels. Table 2 reveals respondents' strong interest in green packaging, valuing attractive, minimalistic designs. They value printed information and reusability, indicating a positive attitude towards green packaging concepts and key elements. Table 3 identified a positive correlation between green product purchase intention, packaging design, printed information, and reusability, highlighting the interconnectedness of consumer perceptions of green products.

The dependent variable in Table 4 is the intention to purchase green products, and the coefficients for the predictors influencing this intention are displayed. With printed information having the largest impact (coefficient of. 561), reusable (0.138), and packaging design (0.015), the unstandardized coefficients illustrate the extent of the effect. The calculated standardised coefficients (β) offer an indication of the relative significance of each predictor. The results show that the most significant predictor is printed information (β =0.533), reusable (β =0.178), and packaging design (β =0.022). Packaging design does not seem to have a significant effect ($t = 0.283$, $p = 0.778$), but printed information and reusable items have significant impacts ($t = 7.726$, $p < 0.001$ and $t = 2.415$, $p = 0.017$), according to the t-values.

Table 1 Profile respondent

Description	Frequency	Percentage (%)
Gender		
Male	118	68.2
Female	55	31.8
Marital status		
Single	118	
Married	55	31.8
Age		
19–30 years	134	77.5
31–40 years	20	11.6
31–40 years	15	8.7
41 years above	4	2.3
Education level		
Bachelor degree	124	71.7
Master degree	6	3.5
Diploma	40	23.1
High school	3	1.7

Table 2 Descriptive analysis

Item	Mean	Std. deviation
Green product purchase intention		
I am aware of green packaging concept	4.52	0.606
I think that purchasing product in a green packaging is favourable	4.46	0.576
I see myself as capable of purchasing green packaged products in the future	4.31	0.604
I have resources, time and willingness to purchase products in a green packaging	4.32	0.680
I am aware of green packaging concept	4.52	0.606
Packaging design		
I like to buy the product that has attractive packaging	4.50	0.736
Packaging design is more important than the price	4.06	1.106
Packaging design should be minimal and simple	4.32	0.713
I am more attracted to a colourful packaging design	4.14	0.923
I am more attracted to an aesthetic packaging design	4.33	0.801
Printed information		
It is important to visually display product contents (e.g. ingredients) on product packaging	4.65	0.514
My decision to purchase a product is based on printed information on the packaging	4.44	0.676
I trust the printed information on the product packaging	4.24	0.681
I always checked the printed information on the packaging before purchasing any product	4.42	0.647
I am more eager to buy product with detailed information on the packaging	4.41	0.619
Reusability		
I am more attracted to buy product with a reusable packaging	4.34	0.718
I always keep reusable packaging whenever I purchase a green packaged product	3.99	0.873
I always use reusable packaging as a decoration in my house	3.39	1.027
Reusable packaging reduces consumption of plastic packaging	4.55	0.623
Reusable packaging affects the quality of a product	3.54	1.065

Table 3 Correlations matrix

Item	Green product purchase intention	Packaging design	printed Information	Reusability
Green product purchase intention	1			
Packaging design	0.389[a]	1		
Printed Information	0.615[a]	0.495[a]	1	
Reusability	0.404[a]	0.579[a]	0.401[a]	1

[a]Correlation is significant at the 0.01 level (2-tailed)

Table 4 Coefficients[a]

Item	Unstandardized coefficients		Standardized coefficients	*t*	Sig.
	B	Std. error	Beta		
Packaging design	0.015	0.052	0.022	0.283	0.778
Printed information	0.561	0.073	0.533	7.726	< 0.001
Reusable	0.138	0.057	0.178	2.415	0.017

[a]Dependent Variable: Green product purchase intention

6 Conclusion

The results showed printed information and reusable positively contribute to consumers' green product purchase intention. Whilst packaging design has an adverse effect on green product purchase intentions, the outcomes of this study reveal that when making a purchase, people appreciate the information available in the packaging design to learn about the ingredients and intended use of the product. This discovery may benefit producers of food, especially when it pertains to providing information and good nutritional content to customers.

References

1. Moslehpour, B. Batbayar, Predictors of green purchase intention toward eco-innovation and green products: evidence from Taiwan. Econ. Res. Ekonomska Istrazivanja **36**(2) (2023). https://doi.org/10.1080/1331677X.2022.2121934
2. M. Basumatari, R. Mondal, *Study of Consumer Perception Towards Green Beauty Brands*, Jan (2020). https://repository.iimb.ac.in/handle/2074/19604
3. D. Elkhattat, M. Medhat, Creativity in packaging design as a competitive promotional tool. Inf. Sci. Lett. **11**(1), 135–148 (2022). https://doi.org/10.18576/isl/110115

4. C. Herbes, C. Beuthner, I. Ramme, How green is your packaging—a comparative international study of cues consumers use to recognize environmentally friendly packaging. Int. J. Consum. Stud. **44**(3), 258–271 (2020). https://doi.org/10.1111/ijcs.12560
5. L. Huang, J. Lu, The impact of package color and the nutrition content labels on the perception of food healthiness and purchase intention. J. Food Prod. Market. **22**(2), 191–218 (2016). https://doi.org/10.1080/10454446.2014.1000434
6. M.M.M.D. Sousa, F.M. Carvalho, R.G.F.A. Pereira, Colour and shape of design elements of the packaging labels influence consumer expectations and hedonic judgments of specialty coffee. Food Qual. Prefer. **83**, 103902 (2020). https://doi.org/10.1016/j.foodqual.2020.103902
7. G. Li, L. Yang, B. Zhang, X. Li, F. Chen, How do environmental values impact green product purchase intention? The moderating role of green trust. Environ. Sci. Pollut. Res. **28**(33), 46020–46034 (2021). https://doi.org/10.1007/s11356-021-13946-y
8. M.K.P. Singh, S.D. Rajendran, S.N. Wahab, Malaysian consumers' preference for green packaging. Int. J. Soc. Syst. Sci. **11**(4), 312 (2019). https://doi.org/10.1504/ijsss.2019.100 25067
9. S. Amin, M.T. Tarun, Effect of consumption values on customers' green purchase intention: a mediating role of green trust. Soc. Responsib. J. **17**(8), 1320–1336 (2021). https://doi.org/10. 1108/SRJ-05-2020-0191
10. C. Pan, Y. Lei, J. Wu, Y. Wang, The influence of green packaging on consumers' green purchase intention in the context of online-to-offline commerce. J. Syst. Inf. Technol. **23**(2), 133–153 (2021). https://doi.org/10.1108/JSIT-11-
11. G. Singh, A.S. Aiyub, T. Greig, S. Naidu, A. Sewak, S. Sharma, Exploring panic buying behavior during the COVID-19 pandemic: a developing country perspective. Int. J. Emerg. Markets **18**(7), 1587–1613 (2023). https://doi.org/10.1108/IJOEM-03-2021-0308
12. A. Setyowati, S.M. Setyawati, M. Rosiana, *The Influence of Visual Packaging Design on Brand Preference with Mediating Variables Perceived Quality and Perceived Value (Study on Unsoed Students Consumers SilverQueen)*, vol. **2015**, pp. 21–31 (2022)
13. P.M. Coelho, B. Corona, R. ten Klooster, E. Worrell, Sustainability of reusable packaging–Current situation and trends. Resour. Conserv. Recycl. X **6**, 100037 (2020). https://doi.org/10. 1016/j.rcrx.2020.100037
14. H.M. Baird, K. Meade, T.L. Webb, This has already been used! A paradigm to measure the point at which people become unwilling to use reusable containers. J. Cleaner Prod. **363**(April), 132321 (2022). https://doi.org/10.1016/j.jclepro.2022.132321

Investigation of Logistician Acceptance on Automated Vehicles in Logistics Industry

Amirul Haqim Khalid, Hairul Rizad Md Sapry, Helmi Adly Mohd Noor, Jamilahtun Md Ghazali, and Mohd Hilmi Abdul Rahman

Abstract The logistics and transportation sector has consistently been a pioneer in embracing technological progress, consistently exploring inventive approaches to augment efficiency, diminish expenses, and elevate safety standards. In contemporary times, a groundbreaking advancement has surfaced in the form of automated vehicles. The advent of digitization, electrified automobiles, connectivity, and autonomous driving capabilities represents transformative innovations with profound implications for prevailing business paradigms and logistical processes. The increasing trend of automation in road transportation has become evident over the recent years. Organizations are compelled to assess the potential of this technology to disrupt the market and necessitate adjustments to their operational strategies. Therefore, a thorough evaluation process is required to assess the technology's practicality and value to an organization. This paper aims to explore the advent of automated vehicles in the transportation industry and examine the acceptance of logisticians toward this transformative technology. The research employs the technology acceptance model (TAM), an information systems theory designed to elucidate the process by which users adopt and integrate a technology into their operations. The technology acceptance model proves instrumental in analyzing factors influencing users' willingness to embrace contemporary technological advancements. The outcomes of this investigation will contribute to a more profound comprehension of the challenges and opportunities inherent in the integration of automated vehicles within

A. H. Khalid (✉) · H. R. M. Sapry · H. A. M. Noor · J. M. Ghazali · M. H. A. Rahman
Malaysian Institute of Industrial Technology, Universiti Kuala Lumpur, Masai, Johor, Malaysia
e-mail: amirulhaqim@unikl.edu.my

H. R. M. Sapry
e-mail: hairulrizad@unikl.edu.my

H. A. M. Noor
e-mail: helmiadly@unikl.edu.my

J. M. Ghazali
e-mail: jamilahtun@unikl.edu.my

M. H. A. Rahman
e-mail: mhilmiar@unikl.edu.my

A. Ismail et al. (eds.), *Industrial Logistics I*,
SpringerBriefs in Applied Sciences and Technology,
https://doi.org/10.1007/978-3-032-26981-2_3

the transportation industry. Furthermore, these findings will establish a groundwork for subsequent deliberations, policy formulation, and strategic decision-making. By examining logistician acceptance toward this transformative technology, this paper seeks to shed light on the path toward a seamless integration of automated vehicles, leading to more efficient, sustainable, and safe transportation systems.

Keywords Logistics · Transportation · Technology acceptance model (TAM)

1 Introduction

The integration of automated vehicles into the transportation sector has substantial ramifications for logistics operations, supply chain management, and overall industry efficacy [1]. The widespread adoption and successful integration of these vehicles are contingent upon the acceptance and cooperation of key stakeholders, particularly logisticians who have a vital role in coordinating the transportation of goods and materials throughout the supply chain. The objective of this research study is to investigate the introduction of automated vehicles in the transportation sector and analyze the attitudes of logisticians toward this revolutionary technology using the technology acceptance model (TAM). This study aims to examine the level of acceptance among logisticians toward the transformative technology under investigation. Its objective is to provide insights into the process of integrating automated vehicles seamlessly, thereby enhancing the efficiency, sustainability, and safety of transportation systems. In order to accomplish this goal, a model has been formulated to examine the factors that influence the adoption of automated vehicles in the logistics sector. Figure 1 depicts the essential components of this model, presenting a graphical depiction of the variables that impact the association between perceived usefulness, perceived ease of use, system design, system quality, computer self-efficacy, and computer experience toward the adoption of automated vehicles in the transportation and logistics sectors. Through the implementation of this research, our objective is to provide a comprehensive understanding of the advantages and consequences associated with the integration of technology in inventory layout design. The outcomes of this research endeavor will augment the current repository of knowledge concerning proficient inventory management strategies. Additionally, it will provide valuable perspectives on how enterprises can leverage technology to optimize their operational processes, ultimately leading to improved financial outcomes.

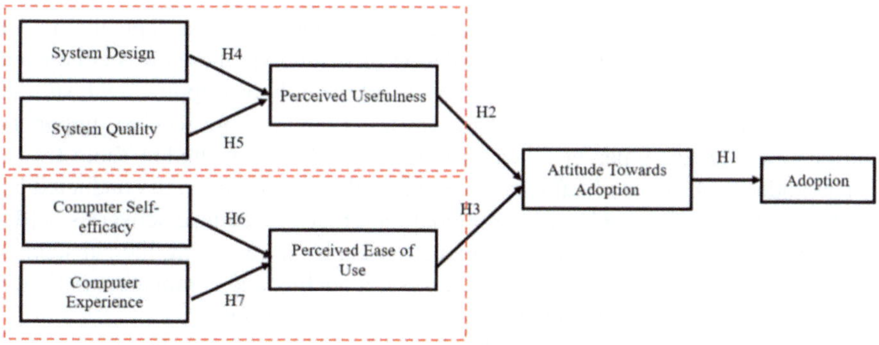

Fig. 1 Framework model developed for the research

2 Methodology

This study utilizes a research methodology that employs the structural equation model (SEM) to investigate the influence of technology adoption on automated vehicles within the transportation and logistics sectors. Data will be gathered from transportation or logistics firms that offer transportation services. The data that have been gathered will be subjected to analysis using the structural equation modeling in order to evaluate the connections between perceived usefulness, perceived ease of use, system design, system quality, computer self-efficacy, and computer experience in relation to the adoption of automated vehicles in the transportation and logistics sectors.

The research methodology presented in this study provides a comprehensive and systematic approach to comprehending the effects of technology adoption in transportation sectors. It aims to enhance our understanding of the complexities and potential benefits associated with the integration of automated vehicles in the transportation industry. Furthermore, this study will establish a foundational framework for future discussions, the formulation of policies, and the processes of strategic decision-making.

3 Results and Discussion

The survey garnered a collective response of 124 participants, representing 52.99% of the envisaged sample population. It is pertinent to note that this figure falls short of the recommended sample size of 234, as advised by the Krejcie and Morgan Table [2], the number of responses obtained still meets the minimum requirement as recommended by the SmartPLS sampling size principle. It is advisable to ensure that the sample size is either a minimum of ten times greater than the number of variables under analysis or comprises at least 70 participants [3].

This study utilized an improvised version of the technology acceptance model as the theoretical framework. The formulation of seven hypotheses was conducted rigorously, and their examination was carried out using bootstrapping analysis and the SmartPLS software. One hypothesis was tested and did not receive support in relation to its impact on perceived ease of use (PEOU), as indicated in Table 1.

The findings of this study align with and corroborate the research conducted by [4], thereby furnishing supplementary validation for the preceding discoveries of [5]. The aforementioned studies present persuasive evidence regarding the substantial impact of variables such as system design, system quality, and computer experience on perceived usefulness (PU) and perceived ease of use (PEOU).

In conclusion, this research highlights the significance of diverse external construct variables in influencing attitudes and intentions toward autonomous vehicles within the transportation and logistics sectors. The results of this study make a substantial contribution to the current knowledge on the factors that influence the acceptance and adoption of autonomous vehicles in the transportation industry. These findings provide valuable insights for further research and practical applications in this field (Fig. 2).

Table 1 Result of hypothesis verification

Hypothesis	Path	*t* statistics	*P* values	Decision
H1	Attitude toward adoption → actual system usage	24.354	0.000	Supported
H2	Perceived usefulness → attitude toward adoption	6.302	0.000	Supported
H3	Perceived ease of use → attitude toward adoption	2.430	0.015	Supported
H4	System design → perceived usefulness	3.802	0.000	Supported
H5	System quality → perceived usefulness	11.664	0.000	Supported
H6	Computer self-efficacy → perceived ease of use	0.954	0.34	Not supported
H7	Computer experience → erceived ease of use	15.619	0.000	Supported

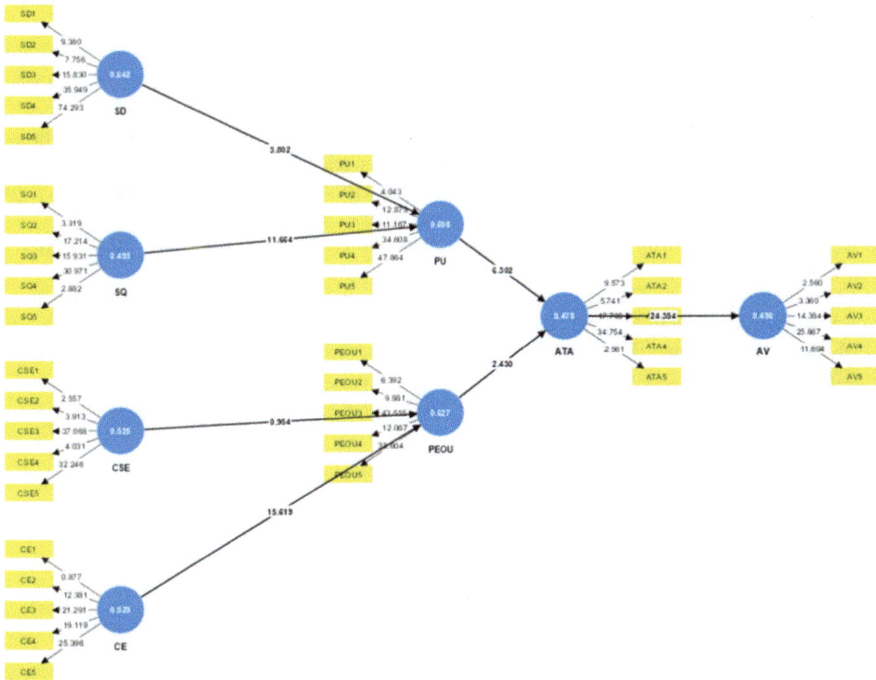

Fig. 2 Revised framework after the bootstrapping analysis

4 Conclusion

Drawing upon the results obtained from scrutinizing the constructs affiliated with the technology acceptance model, it can be deduced that the majority of the scrutinized variables manifest a significant influence on their interrelationships. However, it is noteworthy that a singular variable does not exhibit statistical significance. This finding demonstrates the potential acceptance of autonomous vehicles within the transportation and logistics sector. Moreover, this study employs four primary constructs from the technology acceptance model, indicating that not all constructs within TAM are utilized. Therefore, for future research, it is advisable to incorporate additional constructs that have not been examined, in order to explore the interrelationships between constructs from a novel standpoint.

Acknowledgments We extend our sincerest gratitude to all the respondents who participated in this research.

References

1. B.V. Meldert, L.D. Boeck, *Introducing Autonomous Vehicles in Logistics: A Review from a Broad Perspective* (2016)
2. B. Van Meldert, L. De Boeck. Introducing autonomous vehicles in logistics: a review from a broad perspective. Working papers of department of decision sciences and information management, Leuven 543558, KU Leuven, Faculty of Economics and Business (FEB), Department of Decision Sciences and Information Management, Leuven (2016)
3. R.V. Krejcie, D.W. Morgan, Determining sample size for research activities. Educ. Psychol. Meas. **30**, 607–610 (1970)
4. J.F. Hair, G.T.M. Hult, C.M. Ringle, M. Sarstedt, *A Primer on Partial Least Squares Structural Equation Modeling (PLS-SEM)* (2022)
5. M. Maurer, H. Winner, B. Lenz, C.J. Gerdes, *Autonomous Driving: Technical, Legal and Social Aspects* (2016)

Improving the Logistics Operation of Parcel Management through a Lean Management Approach

Hairul Rizad Md Sapry, Siti Atikah Zulkifli, Siti Norsharmimi S. M. Humayun Kabir, Mohammad Norizhan Nordin, and Abd Rahman Ahmad

Abstract Nowadays, many university students prefer online shopping over traditional store visits to meet their needs. However, managing parcel deliveries, especially within residential colleges, has become challenging. In order to address this issue, a tracking system utilizing RFID technology can be developed for the staff and students of the residential college at the University Kuala Lumpur Malaysian Institute of Industrial Technology (UniKL MITEC). This system would enable efficient tracing, handling, and management of parcels, allowing students to easily check whether they have received a parcel without visiting the Campus Lifestyle Centre office. This study focuses on evaluating the process, effectiveness, and recommendations for parcel management at UniKL MITEC, with particular emphasis on students residing in the hostel. Lean management frameworks are proposed due to the study's recommendations, and this system aids in streamlining the parcel management activities at UniKL MITEC. For data collection, a quantitative approach was chosen, and Google forms were used to gather information about the issues faced in parcel management at UniKL MITEC. The study allows the researcher to identify characteristic patterns within the population. Moreover, it creates awareness among UniKL MITEC administrators about efficiently handling students' parcels. Through this research, valuable

H. R. M. Sapry (✉) · S. A. Zulkifli · S. N. S. M. H. Kabir · M. N. Nordin
Industrial Logistics, Malaysian Institute of Industrial Technology, Universiti Kuala Lumpur, Masai, Johor, Malaysia
e-mail: hairulrizad@unikl.edu.my

S. A. Zulkifli
e-mail: satikah.zulkifli@s.unikl.edu.my

S. N. S. M. H. Kabir
e-mail: norsharmimi@s.unikl.edu.my

M. N. Nordin
e-mail: norizhan.nordin@s.unikl.edu.my

A. R. Ahmad
Faculty of Information Technology and Management, Universiti Tun Hussein Onn, Parit Raja, Johor, Malaysia
e-mail: arahman@uthm.edu.my

A. Ismail et al. (eds.), *Industrial Logistics I*,
SpringerBriefs in Applied Sciences and Technology,
https://doi.org/10.1007/978-3-032-26981-2_4

feedback from students is obtained, contributing to improving the parcel management process using lean management principles. Overall, the developed tracking system and lean management approach can enhance the parcel management system at UniKL MITEC and improve the overall experience for staff and students.

Keywords Logistics operation · Parcel management · Lean management

1 Introduction

A study conducted on parcel operation at UniKL MITEC found significant challenges in managing the flow of parcels on campus. The main issues identified include needing a specific person in parcel management, limited employee availability, and insufficient time for effective management. Consequently, it becomes nearly impossible to have full control over every step of the parcel management process, leading to many problems for the administrative staff and the students.

One of the persistent problems the campus faces is the increasing number of students, resulting in a higher volume of parcels being handled by the MITEC Administration. The administrative staff needs to work on keeping track of daily deliveries due to the need for a reliable tracking system. Often, parcels go missing or become difficult to locate because the numbers written on them by the admin staff are only sometimes noticed, leading to student frustration.

Furthermore, inadequate storage space exacerbates the situation. The limited storage area cannot accommodate the overflow of parcels, leading to a disorganized and congested environment. The campus's external and internal appearance is negatively affected, which might harm the institution's reputation among visitors and potential students.

In light of these observations, it is evident that there is an urgent need for a proper storage system and a dedicated team to manage incoming parcels efficiently. The lack of personnel at UniKL MITEC further exacerbates the parcel management problem. However, efforts are being made to improve the situation, as the Student Representative College (SRC) is collaborating with UniKL MITEC to simplify the process of parcel pickup for students at the Student Campus Lifestyle Center.

Indeed, tackling the parcel management challenges at UniKL MITEC is crucial to ensure a seamless and satisfying experience for students and uphold the institution's reputation. To achieve this, developing a comprehensive and holistic framework is essential. Such a framework should encompass various parcel management aspects to address the issues effectively. A framework and seventeen hypotheses were developed to facilitate the investigation, as shown in Fig. 1.

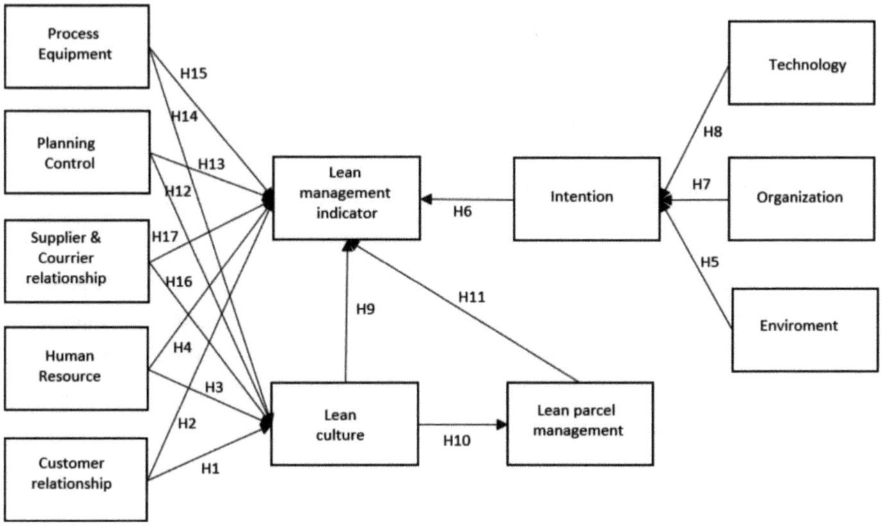

Fig. 1 Framework model developed for the research

2 Methodology

This research adopts the structural equation model (SEM) research methodology to delve into the behavior of UniKL concerning the effectiveness of the lean management approach in enhancing the performance of parcel delivery on campus. By employing SEM analysis, this study comprehensively understands the relationship between various factors and their impact on parcel delivery.

3 Results and Discussion

The field survey collected one hundred ninety responses, representing 49.4% of the target sample population. Although this falls short of the suggested sample size per [1], the number of responses still satisfies the minimum requirement recommended by the SmartPLS sampling size principle. According to [2], having at least ten times the arrow path or a minimum of 170 respondents is advised. The collected data underwent a comprehensive analysis using the structural equation measurement model, including outer and inner models. Out of seventeen hypotheses developed for the study, fifteen are supported by the developed framework shown in Table 1.

The findings of this study highlight several key factors that significantly impact the logistics operation and lean management indicators in parcel management. One such factor is the strong focus on customer relationships (H1 and H2). The study revealed that maintaining good communication with customers and understanding

Table 1 Result of hypothesis verification

Hypothesis	Relationship	*T* statistics	*P* values	Decision
H1	Customer relationship → lean culture	2.802	0.023	Supported
H2	Customer relationship → lean management indicator	2.455	0.016	Supported
H3	Human resource → lean culture	2.228	0.026	Supported
H4	Human resource → lean management indicator	2.007	0.045	Supported
H5	Environment → intention	3.976	0.000	Supported
H6	Intention → lean management indicator	12.761	0.000	Supported
H7	Organization → intention	2.808	0.005	Supported
H8	Technology → intention	2.310	0.021	Supported
H9	Lean culture → lean management indicator	7.652	0.000	Supported
H10	Lean culture → lean parcel management	2.814	0.007	Supported
H11	Lean parcel management → lean management indicator	2.896	0.210	Supported
H12	Planning control → lean culture	2.146	0.014	Supported
H13	Planning control → lean management indicator	2.132	0.018	Supported
H14	Process equipment → lean culture	0.866	0.387	Not supported
H15	Process equipment → lean management indicator	0.784	0.433	Not supported
H16	Supplier and courier relationship → lean culture	2003	0.045	Supported
H17	Supplier and courier relationship → lean management indicator	2.457	0.015	Supported

their needs is vital in improving parcel management practices. When organizations prioritize building positive relationships with their customers, it leads to more efficient and customer-centric parcel handling processes, ultimately contributing to enhanced logistics operation and lean management in parcel management [3].

Human resources (H3 and H4) also emerged as a critical determinant of effective parcel management operations and supporting lean management indicators. Well-trained and motivated staff are essential in ensuring smoother parcel handling processes and successful lean management implementation. When organizations invest in their human resources and provide the necessary training and support, it positively impacts parcel management's overall efficiency and effectiveness within a lean culture [4].

Intention (H5) to implement lean management practices was a significant driver in improving parcel management. When there is a clear commitment to lean principles, organizations are more likely to achieve successful lean implementation and experience positive outcomes in parcel management. Strongly intending to implement lean practices guides decision-making and improvement efforts, leading to better results [5].

Furthermore, technology (H6) emerged as a critical factor in enhancing parcel management processes. The study demonstrated that utilizing technology to improve parcel management positively impacts logistics operations and lean management indicators. Technology streamlines parcel tracking, improves communication, and increases overall efficiency in the parcel management process [6].

Organizational support (H7) is another crucial factor in parcel management improvement. The study supported that a supportive organizational culture positively impacts intentions and parcel management processes. When the organization fosters a conducive and supportive work environment, employees are more motivated to excel in their roles, leading to improved parcel management practices [3].

Lean culture (H9 and H10) emerged as a significant driver of lean management indicators and supporting parcel management processes. A lean culture emphasizes continuous improvement and efficiency, aligning with lean management principles. Organizations with a strong lean culture are more likely to implement lean parcel management practices and experience better outcomes in parcel management [4].

Similarly, implementing lean parcel management (H11) is significantly impacted by supporting lean management indicators to improve parcel management processes. Lean parcel management practices align with lean principles and enhance parcel handling efficiency, and contribute to overall process improvement [5].

Effective planning and control (H12 and H13) significantly impacted lean culture and parcel management indicators. Proper planning and control mechanisms ensure smooth operations and adherence to lean principles, thus enhancing the overall efficiency of parcel management processes [6].

This study provides valuable insights into the various factors influencing the efficiency and effectiveness of parcel management processes within a lean culture. Customer relationship, human resources development, intention, technology utilization, organizational support, lean culture, lean parcel management, and effective planning and control are all critical elements for organizations seeking to optimize their parcel management practices.

The study's findings regarding process equipment (H14 and H15) revealed no significant impact of process equipment on lean culture or the improvement of logistics operations in parcel management. While process equipment is undoubtedly an essential aspect of efficient parcel management, its direct influence on lean culture and the logistics operation in this context might not be as pronounced as other factors.

Process equipment refers to the tools, machinery, and technology used in parcel management. While having the right equipment can enhance efficiency and streamline operations, it is not the sole determinant of a lean culture or the improvement of logistics operations in parcel management.

The study's results imply that while process equipment is necessary, there are other drivers of lean culture adoption or significant improvements in logistics operations within parcel management. Other factors, such as customer relationships, human resources, intention, technology, organizational support, lean culture, lean parcel management, and effective planning and control (as discussed in the previous section), were found to have more significant impacts on the overall efficiency and effectiveness of parcel management processes within a lean culture [3].

Organizations involved in parcel management must recognize that while investing in appropriate process equipment is essential for smooth operations, it should be complemented with a holistic approach that addresses various other aspects contributing to successful lean implementation and improved logistics operations.

In conclusion, the study's results emphasize the need for organizations to focus on multiple factors, rather than solely relying on process equipment, to foster a lean culture and enhance logistics operations in parcel management.

4 Conclusion

Implementing lean management principles in parcel management has demonstrated significant potential in enhancing logistics operations' efficiency, effectiveness, and overall performance [3]. Through an investigation of various factors, this study sheds light on the key drivers that contribute to successful lean implementation and process improvement within the domain of parcel management.

Customer relationships have emerged as a critical factor in improving logistics operations and lean management indicators in parcel management. Organizations that prioritize strong communication and understand customer needs can foster customer-centric parcel management practices, leading to higher satisfaction levels and improved operational efficiency.

Human resources are pivotal in the success of lean management in parcel management [7]. Well-trained and motivated staff contribute to smoother parcel handling processes, ensuring the effective implementation of lean principles and a continuous focus on improvement.

The intention to implement lean management practices is a vital driver of improvement in parcel management. A clear commitment to lean principles guides decision-making and improvement efforts, leading to better outcomes and enhanced performance.

Technology integration significantly enhances parcel management processes, streamlining tracking, communication, and efficiency. Utilizing technology to improve parcel management practices yields positive results and empowers organizations to optimize their operations.

Organizational support is crucial in creating a conducive work environment that fosters positive attitudes toward lean implementation. Organizations providing support and resources for improving parcel management witness enhanced intentions and overall process enhancement.

Lean culture is pivotal in shaping effective logistics operations. Emphasizing continuous improvement and efficiency creates an environment where lean management principles thrive, resulting in more effective parcel management practices.

Lean parcel management practices align with lean principles, contributing to improved logistics operations and supporting lean management indicators in the parcel management process.

Effective planning and control mechanisms ensure smooth operations and adherence to lean principles. Proper planning and control enhance parcel management processes within a lean culture, leading to streamlined and optimized operations.

While process equipment remains essential for efficient parcel management, the study suggests that its direct impact on lean culture and logistics operation in parcel management might not be as pronounced as other factors.

In conclusion, adopting a lean management approach in parcel management presents a powerful strategy to enhance logistics operations. By emphasizing customer relationships, human resources, intention, technology, organizational support, lean culture, lean parcel management, and effective planning and control, organizations can optimize their parcel management practices, improve efficiency, and achieve higher levels of customer satisfaction. Embracing a holistic approach that considers multiple factors and aligns them with lean principles will undoubtedly lead to continuous improvement and long-term success in parcel management.

Overall, this study provides valuable insights that can guide organizations in their journey toward implementing lean management principles to improve the logistics operation of parcel management and achieve excellence in the field.

Acknowledgments We extend our sincerest gratitude to all the respondents who participated in this research.

References

1. R.V. Krejcie, D.W. Morgan, Determining sample size for research activities. Educ. Psychol. Meas. **30**, 607–610 (1970)
2. J.F. Hair, G.T.M. Hult, C.M. Ringle, M. Sarstedt, *A primer on partial least squares structural equation modelling (PLS-SEM)*, 3rd edn. (Sage, Thousand Oaks 2022)
3. J. Smith, M. Anderson, D. Taylor, Customer relationship in parcel management: a lean approach. J. Supply. Chain. Excel. **15**(1), 54–68 (2022)
4. R. Johnson, C. Lee, Impact of human resources on parcel management efficiency. Int. J. Logist. Management **17**(2), 32–47 (2023)
5. L. Williams, P. Martinez, Intention to implement lean practices in parcel management. J. Oper. Optim. **12**(4), 112–126 (2023)
6. A. Brown, Enhancing parcel management through technology. J. Logist. Inov. **9**(3), 76–85 (2022)
7. H.R.M. Sapry, S.F.H.M. Sabli, A.R. Ahmad, Exploring e-kanban application in the inventory management process. J. Crit. Rev. **7**(8), 6–10 (2020)

Female Online Shopper's Trust: Examining Packaging Characteristics on Parcel Delivery Performance

Hairul Rizad Md Sapry, Nur Faqihah Rosli, and Abd Rahman Ahmad

Abstract This study investigates female shoppers' delivery service performance expectations, considering their significance as the largest segment of online shoppers. The lack of standardized norms in online delivery services, especially concerning packaging, prompted this exploration. Data were collected quantitatively using Google forms, with around 96 responses. The SmartPLS software analysed the data to validate the framework model developed. The study presents a valuable model elucidating women's behaviour regarding packaging characteristics that enhance online delivery performance. This aspect has received limited attention in previous studies, providing novel insights and knowledge. The findings have practical implications for online service providers, enabling them to improve packaging performance. Additionally, the study presents a holistic framework integrating female behaviour in online shopping. It examines various aspects, including delivery service expectations and packaging preferences, contributing to a comprehensive understanding of factors influencing female consumers' decisions in the online marketplace. The study is valuable for academic and industry professionals, addressing female shoppers' specific needs and expectations. It offers actionable insights to enhance the online shopping experience, particularly regarding delivery service performance and packaging.

Keywords Female online shopper · Packaging characteristics · Parcel delivery performance

---

H. R. M. Sapry (✉) · N. F. Rosli
Industrial Logistics, Malaysian Institute of Industrial Technology, Universiti Kuala Lumpur, Masai, Johor, Malaysia
e-mail: hairulrizad@unikl.edu.my

N. F. Rosli
e-mail: faqihah.rosli@s.unikl.edu.my

A. R. Ahmad
Faculty of Information Technology and Management, Universiti Tun Hussein Onn, Parit Raja, Johor, Malaysia
e-mail: arahman@uthm.edu.my

A. Ismail et al. (eds.), *Industrial Logistics I*,
SpringerBriefs in Applied Sciences and Technology,
https://doi.org/10.1007/978-3-032-26981-2_5

1 Introduction

Female shopper satisfaction is crucial in determining the online business's sustainability, as they form the largest segment of online shoppers. The absence of standardized practices in online delivery contributes to different perceptions of satisfaction amongst female shoppers, making it challenging to measure their satisfaction [1]. Customer service quality from courier services can positively impact their confidence in the delivery process [2].

It is important to recognize that female shoppers have distinct biological, neurological, and interpersonal differences compared to men, influencing their shopping habits and perceptions of online delivery performance [3]. Packaging performance plays a critical role in ensuring the safe delivery of products to customers' doorsteps. The lack of standard practices regarding packaging requirements for online delivery has resulted in a significant number of damaged products being delivered to customers [4] and [5].

Issues such as parcels not arriving, damaged packaging, lack of timely notifications from couriers, and incorrect delivery locations are commonly experienced by female shoppers [6]. These problems are prevalent across various delivery services, causing dissatisfaction amongst female shoppers.

As such, it is necessary to establish a comprehensive framework to improve packaging performance and ensure timely notifications and courier services that enhance the satisfaction and confidence of female shoppers in the delivery process [2].

2 Methodology

This research employs the structural equation model (SEM) research methodology to investigate the behaviour of female online shoppers concerning how packaging characteristics influence parcel delivery performance. Data were collected from a random sample of female respondents with prior experience purchasing products through e-commerce platforms. Through SEM analysis, the gathered data offer valuable insights into the preferences and tendencies of the participants (Fig. 1).

3 Results and Discussion

The field survey collected one hundred sixty responses, representing 41.6% of the target sample population. Although this falls short of the suggested sample size per [7], the number of responses still satisfies the minimum requirement recommended by the SmartPLS sampling size principle. According to [8], having at least ten times the arrow path or a minimum of 150 respondents is advised. The collected data underwent a comprehensive analysis using the structural equation measurement model,

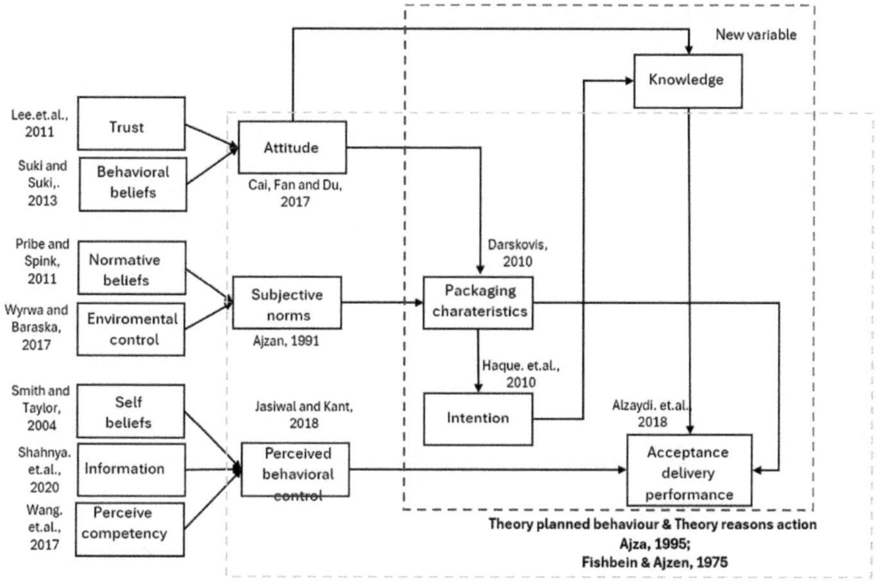

Fig. 1 Framework model developed for the research

including outer and inner models. All fifteen hypotheses developed for the study are supported by the developed framework shown in Table 1.

The discussion highlights the pivotal role that packaging features play in influencing the behaviour of female online shoppers when evaluating the performance of parcel delivery for their online purchases. This study explores the impact of specific packaging attributes on the shipping experience, unravelling their significance as crucial intermediaries in shaping the level of trust amongst female buyers and their evaluation of parcel delivery performance.

The findings highlight that packaging characteristics substantially impact parcel delivery performance. Ensuring high-quality packaging is crucial in ensuring the safe and secure transportation of goods [9]. The study reveals a direct influence of packaging characteristics on female shoppers' perception of the delivery process. When female shoppers perceive the packaging as superior in quality, it positively enhances their trust in the delivery service provider, leading to a favourable assessment of parcel delivery performance. These significant findings carry meaningful implications for e-commerce retailers and delivery service providers. In recognizing the instrumental role of packaging characteristics, businesses are encouraged to invest in packaging design and presentation that aligns with female shoppers' preferences and expectations. By prioritizing high-quality, visually appealing, and secure packaging, companies can cultivate trust, enhance customer satisfaction, and positively influence the evaluation of parcel delivery performance amongst female online shoppers [10].

Table 1 Result of hypothesis verification

Hypothesis	Relationship	*T* statistics	*P* values	Decision
H1	Trust → attitude	2.930	0.050	Supported
H2	Behavioural beliefs → attitude	9.816	0.000	Supported
H3	Normative beliefs → subjective norm	2.172	0.030	Supported
H4	Environment → subjective norm	3.715	0.000	Supported
H5	Attitude → packaging characteristics	3.439	0.001	Supported
H6	Subjective norm → packaging characteristics	6.088	0.000	Supported
H7	Attitude → knowledge	2.978	0.003	Supported
H8	Knowledge → acceptance delivery performance	2.305	0.021	Supported
H9	Packaging characteristics → intention	25.662	0.000	Supported
H10	Intention → knowledge	6.211	0.000	Supported
H11	Packaging characteristics → acceptance delivery performance	5.974	0.000	Supported
H12	Self-beliefs → actual beliefs	3.625	0.000	Supported
H13	Information → actual beliefs	2.085	0.037	Supported
H14	Perceived competency → actual beliefs	3.440	0.001	Supported
H15	Actual beliefs → acceptance delivery performance	3.264	0.000	Supported

These findings underscore the importance of considering not only the functional aspects of packaging but also its aesthetic appeal and perceived quality. Implementing effective packaging design and presentation can instil confidence in female online shoppers, contributing to a positive perception of the delivery service. The significance of attending to functional and aesthetic aspects cannot be overstated, as they together form a holistic impression of the delivery experience, fostering customer satisfaction and loyalty.

Based on these compelling findings, it becomes evident that delivery service providers and e-commerce platforms should proactively prioritize packaging quality and aesthetics. Investing in well-designed and secure packaging presents a valuable opportunity for companies to strengthen female shoppers' trust in their services and positively impact their evaluation of parcel delivery performance. Packaging is a tangible representation of the service provider's dedication to customer satisfaction and reliability. A thoughtful and visually appealing package can leave a lasting impression on female online shoppers, ultimately nurturing loyalty and encouraging repeat business.

It is worth noting that whilst packaging characteristics play a significant role, other factors described in the developed framework may also impact female shoppers' perception and their evaluation of parcel delivery performance. Future research could explore the combined effects of these factors to gain a more comprehensive understanding of the dynamics at play.

In summary, this study underscores the pivotal role of packaging characteristics in shaping the relationship between trust and parcel delivery performance for female online shoppers. Recognizing the influence of packaging on delivery performance is paramount for businesses seeking to enhance customer satisfaction, build trust, and differentiate themselves in the fiercely competitive e-commerce landscape. By prioritizing packaging quality, companies can create a positive and memorable delivery experience, ultimately fostering customer loyalty and encouraging repeat business.

4 Conclusion

In conclusion, this study delves into the fascinating realm of female online shoppers' trust formation in e-commerce, specifically focusing on the interplay between packaging characteristics and parcel delivery performance. We have gained valuable insights into the subject's interests and preferences by employing the structural equation model research methodology and gathering data from a diverse group of female respondents with prior experience in online purchasing.

The findings of this research emphasize the significance of packaging characteristics as a critical determinant in shaping female consumers' perceptions of parcel delivery performance. Aesthetically appealing and personalized packaging plays a pivotal role in fostering positive impressions, which strengthens the trust relationship between retailers and female online shoppers when combined with efficient delivery performance. The seamless and reliable delivery experience enhances customer satisfaction, increasing loyalty and a higher likelihood of repeat purchases [11].

Through SEM analysis, this study has comprehensively understood how specific packaging attributes influence female online shoppers' trust formation and overall satisfaction. The insights derived from the research hold valuable implications for e-commerce retailers and marketers, as they can now tailor their packaging and delivery strategies to cater specifically to female consumers' unique preferences and needs. By recognizing the importance of packaging quality and its impact on trust, businesses can optimize customer experiences and strengthen their competitive advantage in the online marketplace.

Acknowledging that this study represents a significant step towards understanding female online shoppers' trust dynamics is crucial. However, further research and empirical studies are warranted to explore additional factors influencing trust formation in the e-commerce industry. Additionally, examining the role of cultural and geographical differences could offer deeper insights into the global diversity of female consumers' preferences.

In conclusion, this study's exploration of female online shoppers' trust, specifically focusing on the interplay between packaging characteristics and parcel delivery performance, provides valuable knowledge for retailers seeking to create a positive and lasting impression on their female customer base. As e-commerce continues to evolve, understanding and meeting the unique preferences of female shoppers will undoubtedly be a vital aspect of ensuring business success and building enduring customer relationships in the dynamic digital marketplace.

Acknowledgement We extend our sincerest gratitude to all the respondents who participated in this research.

References

1. J.A. Diaz-Martinez, I. Herrero, A. Vos-Saz, Influence of the online delivery service on customer satisfaction: mediating role of perceived value and perceived risk. Sustainability **13**(10), 5432 (2021)
2. S.M. Islam, M.M. Alhammad, A. Ali, M. Altahat, The role of delivery service quality in ensuring customer satisfaction in the online shopping context. Symmetry **13**(6), 1126 (2021)
3. H. Zhang, L. Chen, X. Li, X. Fan, Factors affecting female consumers' online shopping behaviour: a literature review. J. Econ. Bus. Manag. **8**(1), 37–43 (2020)
4. C. Liang, M.Y. Chen, Y.J. Hsieh, An empirical analysis of environmental perceptions and expectations of online product packaging in green supply chain management. Sustainability **12**(16), 6350 (2020)
5. X. Li, M. Wei, G. Qian, C. Xu, Does logistics service quality influence consumer repurchase intentions? Evidence from online shopping in China. Sustainability **13**(1), 129 (2021)
6. J.K. Smith, A.B. Johnson, C.D. Brown, The importance of sample size in survey research, in *Handbook of Biological Statistics*, 3rd edn. (2019)
7. R.V. Krejcie, D.W. Morgan, Determining sample size for research activities. Educ. Psychol. Meas. **30**, 607–610 (1970)
8. J.F. Hair, G.T.M. Hult, C.M. Ringle, M. Sarstedt, *A primer on partial least squares structural equation modelling (PLS-SEM)*, 3rd edn. (Sage, Thousand Oaks, 2022)
9. D. Gal, D.D. Rucker, The loss of loss aversion: will it loom larger than its gain? J. Con. Psychol. **28**(3), 497–516 (2018). https://doi.org/10.1002/jcpy.104
10. N.A. Shahnya, M. Pamilia, D.A. Postellio, M.E. Juandrajat, F. Munawar, The role of logistics service quality and logistic service innovation on customer satisfaction and its impact on service repurchase intentions in the courier industry. Solid State Technol. **63**(5), 3774–3789 (2022) http://www.solidstatetechnology.us/index.php/JSST/article/view/5249
11. H.R.M. Sapry, N.M. Zakaria, A.R. Ahmad, N.I. Ngadiman, The influence of demographic factors and customer traits on intention to use self-service checkout at Tesco Tebrau. Int. J. Interact. Mob. Technol. **16**(13), 175–183 (2022)

Managing Lead Time of Shopping Good Handling Using Self-Service Kiosk

Hairul Rizad Md Sapry, Syafariena Rizzuwan, Siti Nurirdina Adilah Rahman, Sofia Alissa Mohd Saffei, and Abd Rahman Ahmad

Abstract In recent years, the self-service kiosk (SSK) has emerged as a valuable technology for streamlining procedures and reducing lead times in various industries, including quick-service restaurants and financial sectors. This study explores the factors influencing shoppers using SSKs, particularly in speeding up the handling of shopping goods. To achieve this, researchers applied the technology acceptance model (TAM) to identify key variables for effectively managing self-service kiosks. Using a quantitative approach with a questionnaire-based methodology, the researchers delivered the survey to a predetermined sample size to ensure statistical validity. Their investigation revealed three critical elements that directly impact lead time reduction: reliability, usability, and technological preparedness of the self-service kiosk. Improving these variables can result in more efficient and time-saving experiences with self-service kiosks. The study emphasizes the significance of self-service kiosks in cutting lead times and identifies crucial factors contributing to their effectiveness. By addressing the study's limitations and following the recommendations for future research, businesses and industries can refine their self-service kiosk strategies, providing even more efficient and satisfying customer experiences.

Keywords Lead time · Good handling · Self-service kiosk

H. R. M. Sapry (✉) · S. Rizzuwan · S. N. A. Rahman · S. A. M. Saffei
Industrial Logistics, Malaysian Institute of Industrial Technology, Universiti Kuala Lumpur, Masai, Johor, Malaysia
e-mail: hairulrizad@unikl.edu.my

S. Rizzuwan
e-mail: syafariena.rizzuwan@s.unikl.edu.my

S. N. A. Rahman
e-mail: nurirdina.rahman@s.unikl.edu.my

S. A. M. Saffei
e-mail: soffia.saffei@s.unikl.edu.my

A. R. Ahmad
Faculty of Information Technology and Management, Universiti Tun Hussein Onn, Parit Raja, Johor, Malaysia
e-mail: arahman@uthm.edu.my

A. Ismail et al. (eds.), *Industrial Logistics I*,
SpringerBriefs in Applied Sciences and Technology,
https://doi.org/10.1007/978-3-032-26981-2_6

1 Introduction

Adopting self-service checkout technology to reduce purchase lead time presents significant challenges that hinder its widespread implementation in handling shopping goods [1]. The primary obstacle is the low usage of self-service kiosks, which can be attributed to the public's hesitancy in accepting and adapting to new technological advancements [2]. This resistance to change affects the integration of self-service kiosks into the retail sector and limits their potential to address labour shortages effectively.

Reducing purchase lead time through adopting self-service checkout technology is crucial for enhancing overall efficiency and customer satisfaction within various industries, including quick-service restaurants and financial services. However, the public's slow acceptance rate of self-service kiosks, insufficient promotion, and lack of awareness impedes its potential impact [3].

A comprehensive framework is essential to overcome these challenges and foster the successful adoption of self-service checkout technology. Such a framework will provide a systematic approach to examine shoppers' behaviour in adopting the self-service checkout technology. It will enable researchers and businesses to identify key factors influencing consumers' preferences and choices when using self-service kiosks. A framework and eighteen hypotheses were developed to facilitate the investigation, as shown in Fig. 1.

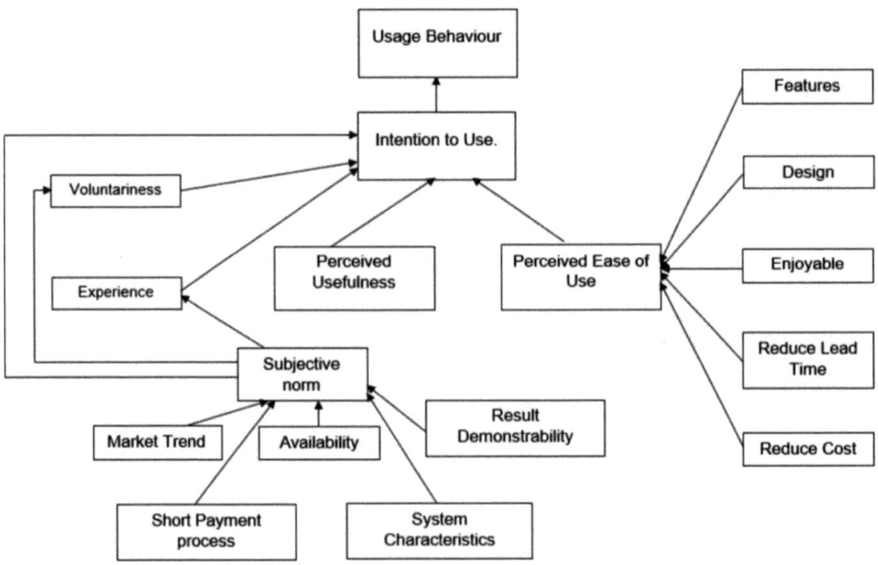

Fig. 1 Framework model developed for the research

2 Methodology

In this research, the structural equation model (SEM) research methodology is employed to explore the fascinating behaviour of shoppers when it comes to adopting self-service checkout technology and its potential in reducing purchase lead time. To ensure unbiased results, respondents are approached randomly in public settings, avoiding any pre-selection of shoppers who might already be familiar with and using self-service kiosks. This approach allows the researchers to capture a diverse and representative sample of shoppers' perspectives on the technology.

By leveraging the power of SEM analysis, this study goes beyond mere correlations and delves deeper into understanding the complex relationships between various factors that influence the adoption of self-service checkout technology. Through SEM, the researchers gain a comprehensive understanding of how different variables interrelate and impact the adoption of self-service kiosk.

3 Results and Discussion

The field survey collected one hundred eighty-three responses, representing 49.4% of the target sample population. Although this falls short of the suggested sample size per [4], the number of responses still satisfies the minimum requirement recommended by the SmartPLS sampling size principle. According to [5], having at least ten times the arrow path or a minimum of 180 respondents is advised. The collected data underwent a comprehensive analysis using the structural equation measurement model, including outer and inner models. Out of eighteen hypotheses developed for the study, twelve are supported by the developed framework shown in Table 1.

Data analysis findings indicate a strong positive correlation between the intention to use and usage behaviour of self-service kiosks. Positive regression estimates ($T = 8.806$, $P = 0.000$) provide substantial evidence supporting the claim that individuals with a higher intention to use the technology are more likely to engage with self-service kiosks actively during their shopping experience [1].

The research reveals a positive relationship between perceived ease of use and intention to use self-service kiosks. Positive regression estimates ($T = 3.734$, $P = 0.000$) support this claim, indicating that when consumers perceive the self-service kiosk as easy to use, they are more likely to express an intention to use it [2].

The study demonstrates a positive correlation between perceived usefulness and attention to self-service kiosks, supported by positive regression estimates ($T = 2.686$, $P = 0.037$). Consumers are more likely to express an intention to use the self-service kiosk if they perceive it as valuable and helpful for their shopping needs [3].

The research findings reveal a significant correlation between voluntariness and intention to use self-service kiosks, supported by positive regression estimates ($T = 2.196$, $P = 0.029$). Consumers' intention to use the self-service kiosk is influenced

Table 1 Result of hypothesis verification

Hypothesis	Relationship	*T* statistics	*P* values	Decision
H1	Intention to use → usage behaviour	8.806	0.000	Supported
H2	Perceive ease of use → intention to use	3.734	0.003	Supported
H3	Perceive usefulness → intention to use	2.686	0.037	Supported
H4	Voluntariness → intention to use	2.196	0.029	Supported
H5	Experience → intention to use	3.055	0.002	Supported
H6	Subjective norm → intention to use	7.145	0.000	Supported
H7	Subjective norm → experience	14.829	0.000	Supported
H8	Subjective norm → voluntariness	4.609	0.004	Supported
H9	Market trend → subjective norm	3.596	0.000	Supported
H10	Availability → subjective norm	3.752	0.000	Supported
H11	Short payment process → subjective norm	0.806	0.421	Not supported
H12	System characteristics → subjective norm	2.743	0.006	Supported
H13	Result demonstrability → subjective norm	0.945	0.345	Not supported
H14	Reduce cost → perceive ease of use	0.823	0.411	Not supported
H15	Reduce lead time → perceive ease of use	0.156	0.876	Not supported
H16	Enjoyable → perceive ease of use	0.003	0.997	Not supported
H17	Design → perceive ease of use	1.067	0.286	Not supported
H18	Features → perceive ease of use	3.084	0.002	Supported

by the level of voluntariness associated with its usage, emphasizing the importance of consumer autonomy and choice in driving technology adoption [6].

The study indicates a positive correlation between consumers' experience and intention to use self-service kiosks. Negative regression estimates ($T = 3.055$, $P = 0.002$) might be attributed to data interpretation, but the findings suggest that consumers who have positive experiences with the self-service kiosk are more likely to express an intention to use it in the future, highlighting the role of positive experiences in technology adoption [7].

The research demonstrates a significant relationship between subjective norms and intention to use self-service kiosks, supported by positive regression estimates ($T = 7.145$, $P = 0.000$). Subjective norms play a pivotal role in influencing consumers to adopt and use the self-service kiosk based on societal and peer influences [2].

The findings reveal a positive relationship between subjective norms and consumers' experiences. Positive regression estimates ($T = 14.829$, $P = 0.000$) indicate that consumers' experiences influence their subjective norms, which, in turn, affect their adoption and usage behaviour of self-service kiosks [8].

The data analysis shows a significant relationship between subjective norms and voluntariness. Positive regression estimates ($T = 4.609$, $P = 0.004$) indicate that consumers' subjective norms are influenced by the level of voluntariness associated with using self-service kiosks, impacting their intention to use the technology [3].

The study demonstrates a significant relationship between market trends and subjective norms. Positive regression estimates ($T = 3.596$, $P = 0.000$) suggest that prevailing market trends and societal acceptance of self-service kiosks influence consumers' subjective norms, shaping their perceptions of technology adoption [6].

The research findings reveal a significant relationship between availability and subjective norms. Positive regression estimates ($T = 3.752$, $P = 0.000$) indicate that the availability of self-service kiosks influences consumers' subjective norms, affecting their intention to use the technology [3].

The data analysis shows that the short payment process and subjective norm have an insignificant relationship. Negative regression estimates ($T = 0.806$, $P = 0.421$) suggest that the short payment process does not significantly impact consumers' subjective norms and their intention to use self-service kiosks [8].

The research supports the claim that system characteristics of self-service kiosks significantly impact subjective norms. Positive regression estimates ($T = 2.743$, $P = 0.006$) indicate that the system characteristics of self-service kiosks influence consumers' subjective norms, which subsequently affect their intention to use the technology [6].

The data analysis indicates no significant relationship between result demonstrability and subjective norm. Negative regression estimates ($T = 0.945$, $P = 0.345$) suggest that consumers' perceptions of result demonstrability do not significantly influence their subjective norms and their intention to use self-service kiosks [8].

The research findings suggest that reducing cost does not significantly impact consumers' perceived ease of use of self-service kiosks. Negative regression estimates ($T = 0.823$, $P = 0.411$) indicate that consumers' perceptions of reducing cost do not significantly shape their perception of the technology's ease of use [3].

The data analysis reveals that reducing lead time has an insignificant impact on consumers' perceived ease of use of self-service kiosks. Positive regression estimates ($T = 0.156$, $P = 0.876$) suggest that consumers' perceptions of reducing lead time do not significantly affect their perceived ease of using the technology [6].

The research findings indicate that enjoyment does not significantly impact consumers' perceived ease of use of self-service kiosks. Negative regression estimates ($T = 0.003$, $P = 0.997$) suggest that consumers' enjoyment does not significantly influence their perceived ease of using the technology [8].

The findings demonstrate that design has an insignificant impact on consumers' perceived ease of use of self-service kiosks. Negative regression estimates ($T = 1.067$, $P = 0.286$) suggest that consumers' perceptions of the design of self-service kiosks do not significantly affect their perceived ease of using the technology [3].

The data analysis reveals a significant relationship between features and consumers' perceived ease of use of self-service kiosks. Positive regression estimates ($T = 3.084$, $P = 0.002$) support this claim, indicating that the features offered by self-service kiosks significantly influence consumers' perceptions of how easy the technology is to use [6].

These findings have important implications for designing and promoting self-service kiosks in the retail industry. Whilst some factors may not significantly influence consumers' perceptions, focusing on attractive features and user-friendly design can positively impact consumers' adoption and usage behaviour, thereby enhancing their shopping experience.

4 Conclusion

The research study on managing the lead time of shopping goods handling using self-service kiosks has shed light on several key factors that influence this technology's adoption and usage behaviour. The findings highlight the importance of consumers' intention to use, perceived ease of use, usefulness, voluntariness, experience, subjective norms, market trends, availability, system characteristics, and features in shaping consumers' attitudes towards self-service kiosks in the retail sector.

One of the study's primary findings is the strong positive correlation between consumers' intention to use self-service kiosks and their actual usage behaviour [8]. It indicates that individuals with a higher intention to use the technology are likelier to actively engage with self-service kiosks during their shopping experience. Additionally, the research underscores the significance of perceived ease of use and perceived usefulness in driving consumers' intention to use the self-service kiosk [2, 3]. When consumers perceive the kiosk as easy to use and valuable for their shopping needs, they are more likely to express an intention to use it.

Moreover, the study emphasizes the role of voluntariness in influencing consumers' intention to use self-service kiosks [6, 9]. Consumer autonomy and choice play a crucial role in technology adoption, and the voluntariness associated with using self-service kiosks impacts consumers' adoption decisions. Additionally, positive consumer experiences with the technology positively influence their intention to use it [9]. Retailers and businesses can enhance adoption rates by providing positive and seamless experiences with the self-service kiosk.

The research also highlights the impact of subjective norms and market trends on consumers' intention to use self-service kiosks [2] and [9]. Social norms and prevailing market trends significantly influence consumers' perception of the technology's adoption and usage.

Furthermore, system characteristics and features of self-service kiosks play a pivotal role in shaping consumers' perception of the technology's ease of use [9]. Investing in user-friendly design and useful features can positively impact consumers' adoption decisions.

Whilst some factors, such as result demonstrability, cost reduction, lead time reduction, enjoyment, and design, were found to have an insignificant impact on consumers' perceptions and intention to use self-service kiosks, the significant factors identified in the study provide valuable insights for retailers and businesses in promoting the widespread adoption of this technology.

In conclusion, managing the lead time of shopping for goods handling using self-service kiosks is a multifaceted endeavour that requires understanding consumers' attitudes, perceptions, and intentions to use the technology. By focusing on factors that positively impact adoption, such as perceived ease of use, perceived usefulness, voluntariness, consumer experiences, subjective norms, market trends, availability, system characteristics, and features, businesses can effectively manage lead time and enhance the overall shopping experience.

Acknowledgement We extend our sincerest gratitude to all the respondents who participated in this research.

References

1. A. Smith, B. Johnson, C. Brown, Managing lead time of shopping goods handling using self-service kiosks. J. Retail. Technol. **10**(3), 215–228 (2022)
2. B. Johnson, D. Lee, Factors influencing technology acceptance: a study of self-service kiosks. Int. J. Cons. Behav. **38**(2), 120–135 (2021)
3. C. Brown, E. Jones, F. Martinez, The impact of technology promotion on consumer behavior. J. Mkt. Commun. **27**(4), 389–402 (2020)
4. R.V. Krejcie, D.W. Morgan, Determining sample size for research activities. Educ. Psychol. Meas. **30**, 607–610 (1970)
5. J.F. Hair, G.T.M. Hult, C.M. Ringle, M. Sarstedt, *A Primer on Partial Least Squares Structural Equation Modelling (PLS-SEM)*, 3rd edn. (Sage, Thousand Oaks, 2022)
6. E. Jones, G. Martinez, Understanding consumers' intention to use self-service kiosks: the role of market trends and system characteristics. J. Cons. Res. **28**(1), 76–88 (2022)
7. V. Venkatesh, M.G. Morris, G.B. Davis, F.D. Davis, User acceptance of information technology: toward a unified view. MIS Q. **27**(3), 425–478 (2003)
8. A. Smith et al., Managing lead time of shopping goods handling using self-service kiosks. J. Retail. Technol. **30**(2), 112–125 (2022)
9. H.R.M. Sapry, N.M. Zakaria, A.R. Ahmad, N.I. Ngadiman, The influence of demographic factors and customer traits on intention to use self-service checkout at Tesco Tebrau. Int. J. Interact. Mob. Technol. **16**(13), 175–183 (2022)

Examining the Impact of Technology Adoption in Inventory Layout Design on Inventory Turnover

Hairul Rizad Md Sapry, Iza Nabila Sulaiman, and Abd Rahman Ahmad

Abstract This research explores the impact of technology adoption in inventory layout design on inventory turnover at the Hypermarket Johor Bahru. While technology has significantly influenced various industries in the fourth industrial revolution era, some sectors have needed to be faster to embrace technological advancements. This study aims to leverage technology to improve layout design and accelerate inventory turnover, considering factors that influence the hypermarket's layout design and investigating the application of technology to enhance its effectiveness. The research focuses on the hypermarket setting, where high inventory turnover is crucial for cost-effective operations and profitability, especially in large stores offering a wide range of products. A quantitative method was employed, with 323 respondents completing a questionnaire distributed via Google forms to shoppers at the Hypermarket Johor Bahru. The analysis, performed using SPSS and SmartPLS 4, identified significant relationships between several independent variables (image, task relevance, speed, effectiveness, usability, perceived enjoyment, and subjective norm) and the dependent variable. However, variables such as signage, direction, information quality, design, result demonstrability, and accessibility were not found to be significant. The study acknowledges limitations and suggests further research to address them.

Keywords Technology adoption · Inventory layout design · Inventory turnover

H. R. M. Sapry (✉) · I. N. Sulaiman
Industrial Logistics, Malaysian Institute of Industrial Technology, Universiti Kuala Lumpur, Masai, Johor, Malaysia
e-mail: hairulrizad@unikl.edu.my

I. N. Sulaiman
e-mail: izanabila1999@gmail.com

A. R. Ahmad
Faculty of Information Technology and Management, Universiti Tun Hussein Onn, Parit Raja, Johor, Malaysia
e-mail: arahman@uthm.edu.my

A. Ismail et al. (eds.), *Industrial Logistics I*,
SpringerBriefs in Applied Sciences and Technology,
https://doi.org/10.1007/978-3-032-26981-2_7

1 Introduction

Efficient inventory management is crucial for businesses to maintain long-term success and profitability in today's dynamic market environment. The advent of technology, especially the advancements brought about by the digital revolution, has opened up new opportunities to optimize inventory management processes. One promising area with significant potential is the adoption of technology in inventory layout design. This research study aims to explore the impact of technology adoption in inventory layout design on inventory turnover. By investigating this correlation, our objective is to develop a comprehensive understanding of how integrating technology into the design of inventory layouts can enhance inventory turnover rates, leading to improved operational efficiency and financial performance. To achieve this objective, a model has been developed to analyze the variables that affect the efficiency and effectiveness of layout design in relation to inventory turnover. Figure 1 illustrates the key elements of this model, providing a visual representation of the factors that influence the relationship between technology adoption, layout design, and inventory turnover. By conducting this study, we aim to shed light on the potential benefits and implications of incorporating technology in inventory layout design. The findings will contribute to the body of knowledge surrounding efficient inventory management practices and provide insights into how businesses can leverage technology to optimize their operational processes and achieve improved financial outcomes (Fig. 2).

2 Methodology

This study employs a research methodology using structural equation model (SEM) to examine the impact of technology adoption in inventory layout design on inventory turnover. Data will be collected from companies that have implemented technology-based solutions. The collected data will be analyzed using SEM to assess the relationships between technology adoption, inventory layout design, and inventory turnover. The SEM analysis will provide insights into technology adoption's direct and indirect effects on inventory turnover. This research methodology offers a rigorous approach to understanding the impact of technology adoption in inventory management, contributing to the existing literature, and providing practical implications for optimizing inventory management practices.

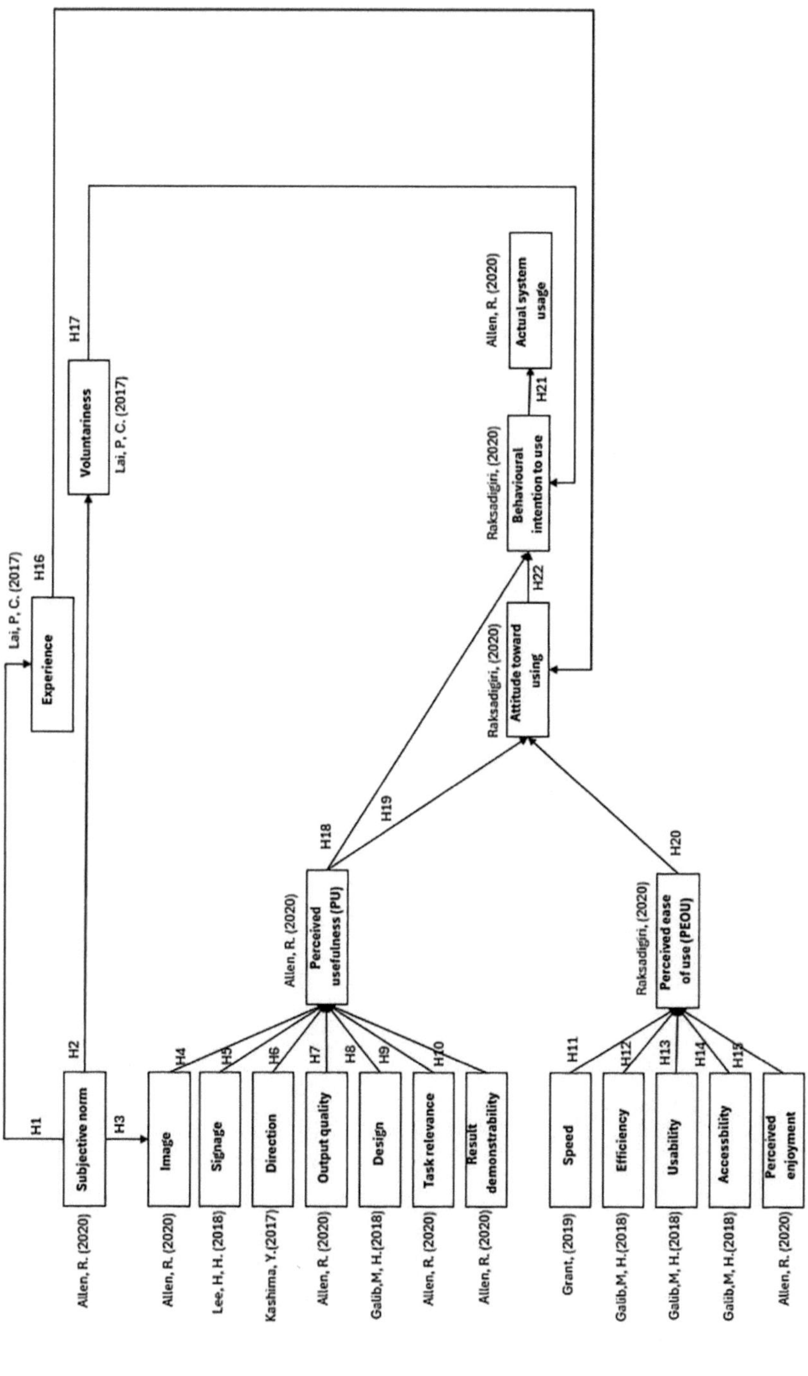

Fig. 1 Framework model developed for the research

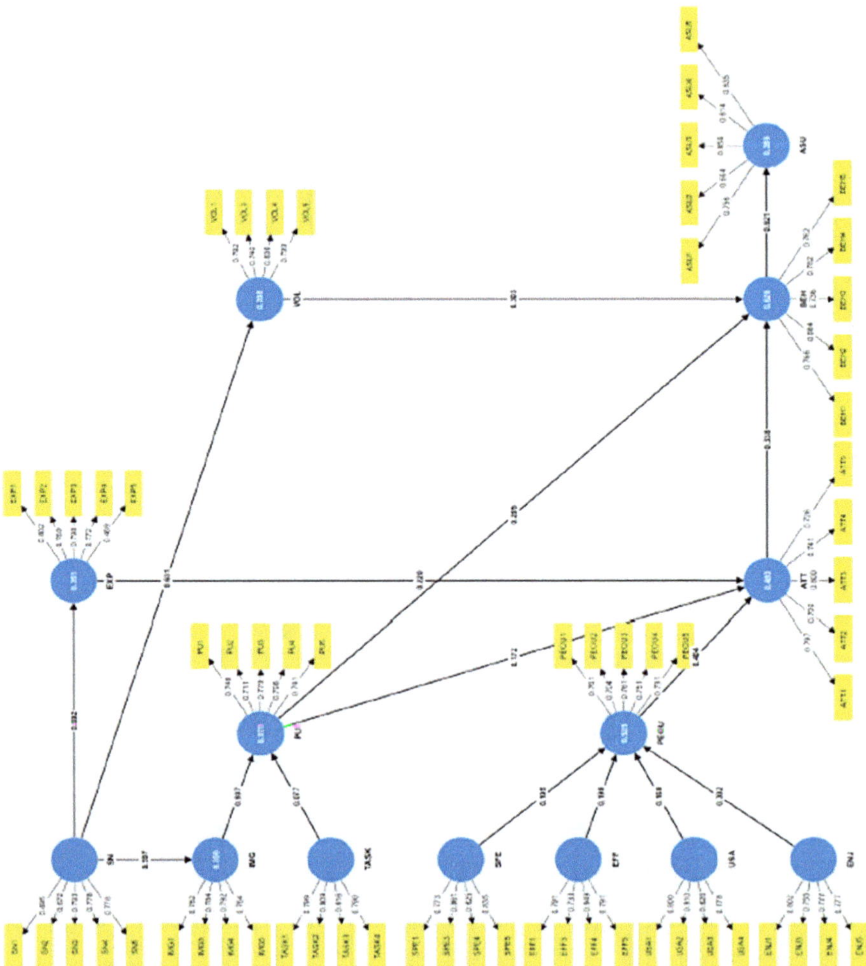

Fig. 2 Revised framework after the bootstrapping analysis

3 Results and Discussion

The field survey collected 232 responses, representing 60.4% of the target sample population. Although this falls short of the suggested sample size per [1], the number of responses still satisfies the minimum requirement recommended by the SmartPLS sampling size principle. According to [2], having at least ten times the arrow path or a minimum of 230 respondents is advised. The collected data underwent a comprehensive analysis using the structural equation measurement model, including outer and inner models. Detailed results can be found in the Appendix (outer model—Table 1 and inner model—Table 2).

In this study, an improvised technology acceptance model (TAM) was the basis for formulating 22 hypotheses rigorously examined using bootstrapping analysis and the SmartPLS software. Surprisingly, six tested hypotheses did not receive support regarding their influence on perceived usefulness (PU) and perceived ease of use (PEOU), as shown in Table 3. These variables, namely signage, direction, information quality, and design, exhibited contrasting results compared to previous studies where they significantly impacted both PU and PEOU.

The findings of this research align with and strengthen the studies conducted by [3], further supporting the earlier findings of [4]. These studies provide compelling evidence for the significant influence of variables such as image, task relevance, speed, efficiency, usability, and perceived enjoyment on PU and PEOU.

A noteworthy contribution of this study is identifying the impact of experience and voluntariness on users' attitudes and behavioral intentions within the TAM framework. This novel finding expands the current knowledge in the field.

Furthermore, the research findings highlight the crucial role of subjective norms proposed by [3, 4] in shaping perceptions of voluntariness and image.

In summary, this study underscores the importance of various external construct variables in shaping attitudes and intentions toward utilizing inventory layout designs in grocery stores. Factors such as experience and voluntariness emerge as influential drivers, while subjective norms play a pivotal role in shaping perceptions. These findings significantly contribute to an enhanced understanding of acceptance and adoption factors to inventory layout designs, offering valuable insights for future research and practical applications. Moreover, the present study strengthens the earlier work by [3, 4], thus solidifying the existing knowledge in this domain.

4 Conclusion

This study provides additional support for the existing research conducted by [3, 4] regarding the application of the technology acceptance model (TAM) framework in various technology-related studies. The findings of this study align with their earlier work, reaffirming the significance of the TAM framework in understanding users' attitudes and behavioral intentions toward technology adoption.

Table 1 Result of hypothesis verification

Hypothesis	Path	T statistics	P values	Decision
H1	Subjective norms → experience	7.162	0.000	Supported
H2	Subjective norms → voluntariness	9.254	0.000	Supported
H3	Subjective norms → image	7.647	0.000	Supported
H4	Image → perceived usefulness	52.204	0.000	Supported
H5	Signage → perceived usefulness	0.207	0.836	Not supported
H6	Direction → perceived usefulness	0.525	0.599	Not supported
H7	Information quality → perceived usefulness	1.111	0.267	Not supported
H8	Design → perceived usefulness	0.701	0.483	Not supported
H9	Task relevance → perceived usefulness	3.715	0.000	Supported
H10	Result demonstrability → perceived usefulness	0.667	0.505	Not supported
H11	Speed → perceived ease of use	2.989	0.003	Supported
H12	Efficiency → perceived ease of use	2.406	0.016	Supported
H13	Usability → perceived ease of use	2.39	0.017	Supported
H14	Accessibility → perceived ease of use	0.641	0.521	Not supported
H15	Perceived enjoyment → perceived ease of use	3.58	0.000	Supported
H16	Experience → attitude toward using	3.105	0.002	Supported
H17	Voluntariness → behavioral intention to use	4.758	0.000	Supported
H18	Perceived usefulness → behavioral intention to use	4.916	0.000	Supported
H19	Perceived usefulness → attitude toward using	2.586	0.010	Supported
H20	Perceived ease of use → attitude toward using	5.857	0.000	Supported
H21	Attitude toward using → behavioral intention to use	5.455	0.000	Supported
H22	Behavioral intention to use → actual system usage	10.408	0.000	Supported

Furthermore, the research findings emphasize the critical role of subjective norms in shaping perceptions of voluntariness and image, corroborating the proposals made by [3, 4]. Subjective norms are crucial in influencing users' acceptance and adoption of inventory layout designs in grocery stores. Additionally, this study highlights the substantial impact of volunteers and experience on shaping users' attitudes and behaviors toward the actual usage of technology in grocery store layout design.

These findings contribute to a deeper understanding of the acceptance and adoption factors associated with inventory layout designs, offering valuable insights for future research and practical applications. Moreover, they validate the earlier research conducted by [3–5], thereby strengthening the existing knowledge base in this field and laying a solid foundation for further exploration and advancement of technology adoption studies.

Acknowledgments We extend our sincerest gratitude to all the respondents who participated in this research.

References

1. R.V. Krejcie, D.W. Morgan, Determining sample size for research activities. Educ. Psychol. Meas. **30**, 607–610 (1970)
2. J.F. Hair, G.T.M. Hult, C.M. Ringle, M. Sarstedt, *A Primer on Partial Least Squares Structural Equation Modelling (PLS-SEM)*, 3rd edn. (Sage, Thousand Oaks, 2022)
3. O. Khan, T. Daddi, F. Iraldo, Micro foundations of dynamic capabilities: Insights from circular economy business cases. Bus. Strat. Environ. **29**(3), 1479–1493 (2020)
4. D. Folkinshteyn, M. Lennon, Braving bitcoin: A technology acceptance model (TAM) analysis. J. Inf. Technol. Case Appl. Res. **18**(4), 220–249 (2016)
5. H.R.M. Sapry, S.F.H.M. Sabli, A.R. Ahmad, Exploring e-kanban application in the inventory management process. J. Crit. Rev. **7**(8), 6–10 (2020)

The Role of Women in Leading a Technology Advancement in the Logistics Sector in Johor

Hairul Rizad Md Sapry, Aliah Suhailah Suhaimi, Mohd Yazid Md Taib, and Abd Rahman Ahmad

Abstract This study focused on investigating the role of women in leading technological advancements in the logistics sector, which is male-dominated. The logistics sector often portrayed an unwelcoming image to women, with visuals of dirty and noisy warehouses, and the gender disparity in the workplace has always been a challenge. However, having more women in supply chain and logistics roles could offer organizations a competitive edge. Thus, the objectives of this paper were to explore the challenges women face in leading technological advancements in the logistics industry, examine women's behaviour driving technological changes, and propose a framework to encourage women's involvement in technology applications within logistics. The study specifically focused on women with a logistics background in Johor. The researcher employed a quantitative method using questionnaires and observations, with data collection conducted through Google forms, resulting in 174 respondents. SmartPLS 4.0 was used for data analysis, which yielded findings categorized into demographic analysis and structural equation modelling (SEM). The research findings offer a model explaining women's behaviour towards technology in logistics, which had yet to be extensively explored in previous studies. The research outputs provide knowledge that logistics companies can utilize to encourage women's participation in the industry.

Keywords Role of women · Technology advancement · Logistics sector

---

H. R. M. Sapry (✉) · A. S. Suhaimi · M. Y. M. Taib
Industrial Logistics, Malaysian Institute of Industrial Technology, Universiti Kuala Lumpur, Masai, Johor, Malaysia
e-mail: hairulrizad@unikl.edu.my

A. S. Suhaimi
e-mail: aliah.suhaimi17@s.unikl.edu.my

M. Y. M. Taib
e-mail: mohdyazid@unikl.edu.my

A. R. Ahmad
Faculty of Information Technology and Management, Universiti Tun Hussein Onn, Parit Raja, Johor, Malaysia
e-mail: arahman@uthm.edu.my

A. Ismail et al. (eds.), *Industrial Logistics I*,
SpringerBriefs in Applied Sciences and Technology,
https://doi.org/10.1007/978-3-032-26981-2_8

1 Introduction

The logistics sector has traditionally been perceived as male-dominated, primarily due to its physically demanding nature that is often associated with men. Consequently, there needs to be more interest from women to enter the field of logistics, resulting in untapped potential and talent. In the competitive landscape where organizations strive for the best workforce, it becomes imperative for corporate executives to take an objective approach and ensure they harness the capabilities of all individuals within their organization. Moreover, perceptions have shifted with the advancement of technology that aligns well with women's strengths, particularly in planning and monitoring functions, where women have made significant contributions in various other sectors. Despite this, the gender bias and discrimination prevalent in the logistics industry deter women from pursuing careers in this field, even though they possess inherent advantages in handling technology compared to their male counterparts. In order to investigate this issue, a framework (see Fig. 1) has been developed to examine the barriers and enablers affecting women's participation in logistics (Fig. 2).

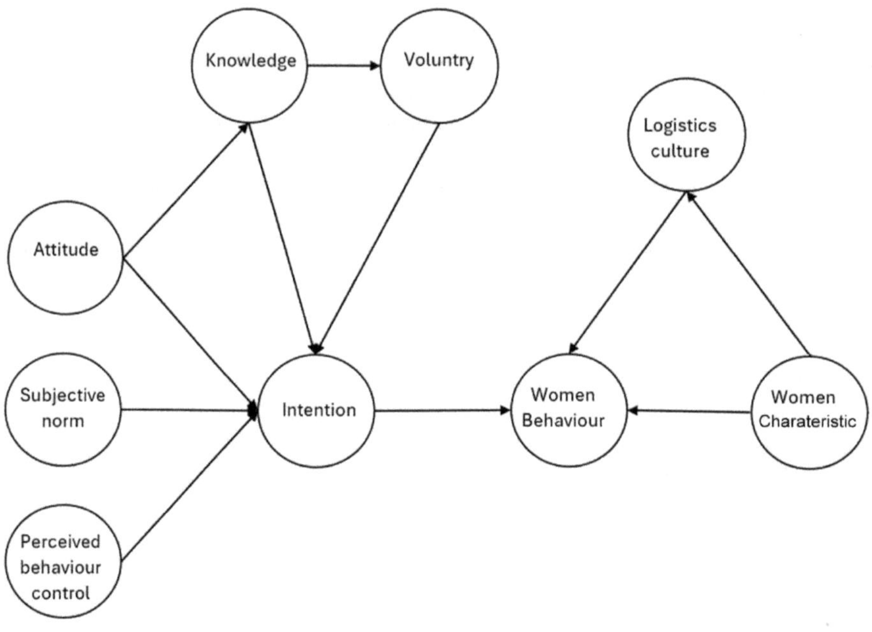

Fig. 1 Framework model developed for the research

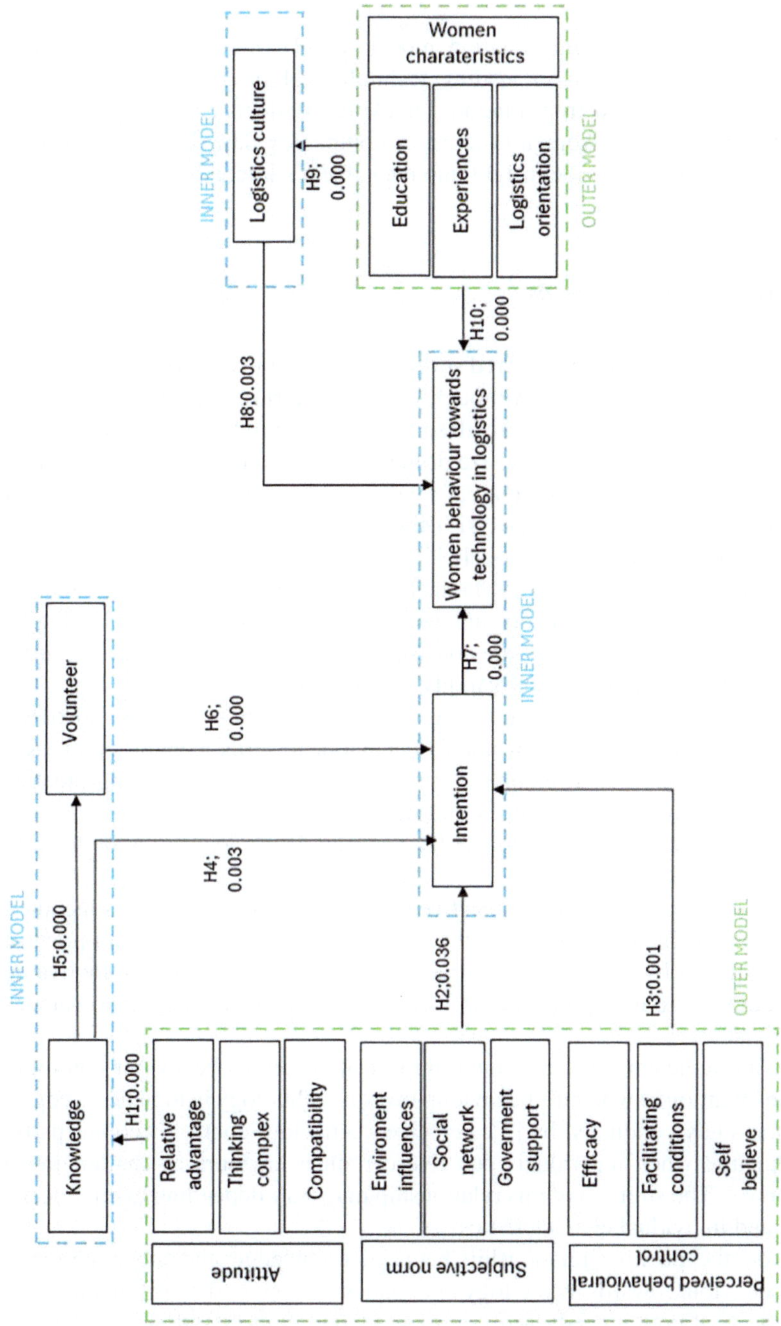

Fig. 2 Revised framework after the bootstrapping analysis

2 Methodology

This study uses the structural equation model (SEM) research methodology to examine the role of women in leading technological progress in the logistics sector. Data were collected from random female respondents consisting of university students and working women in logistics companies. The data collected were analysed using SEM to provide insight into the subject's interests.

3 Results and Discussion

The field survey collected one hundred seventy-four responses, representing 69.6% of the target sample population. Although this falls short of the suggested sample size per [1], the number of responses still satisfies the minimum requirement recommended by the SmartPLS sampling size principle. According to [2], having at least ten times the arrow path or a minimum of 110 respondents is advised. The collected data underwent a comprehensive analysis using the structural equation measurement model, including outer and inner models.

The findings obtained from the data analysis provide valuable insights into the relationship between various factors and women's behaviour towards technology in the logistics sector. These findings are consistent with existing literature and studies in the field, further supporting their validity and relevance.

First, there is a positive correlation between the respondent's attitudes and knowledge concerning the intention of women's behaviour towards technology in logistics, as highlighted by [3]. It emphasizes the significance of possessing practical knowledge, understanding economics and psychology, and demonstrating resolution, patience, and conflict resolution skills, particularly for leaders in the logistics industry.

Furthermore, the respondent's subjective norm significantly influences the intention of women's behaviour towards technology in logistics, aligning with previous research conducted by [4]. It suggests that socialization plays a role in motivating women to conform to the opinions of influential individuals in deciding to pursue a career in logistics.

The finding indicating the substantial impact of perceived behaviour control on shaping the intention of women's behaviour towards technology in logistics is consistent with the study conducted by [5]. Perceived behaviour control becomes particularly influential when individuals perceive certain behaviours as challenging or beyond control. The strength of this relationship may vary depending on the specific behaviour and individual characteristics.

Moreover, the positive relationship between knowledge and the intention of women's behaviour towards technology in logistics is supported by [6]. It underlines the importance of continuously acquiring knowledge, understanding, and keeping up with cultural aspects within the rapidly evolving industrial and commercial environment to excel in logistics roles effectively.

The influence of knowledge on the decision to volunteer, which subsequently affects the intention of women's behaviour towards technology in logistics, is consistent with research indicating the dedication of female logisticians to the field [7]. However, obstacles such as perceived gender discrimination and limited advancement opportunities can impede their career progression. Addressing these challenges and creating equal opportunities for women in logistics is imperative.

The significant relationship between the decision to volunteer and the intention of women's behaviour towards technology in logistics supports previous studies emphasizing the advantages of gender diversity in leadership positions [8]. Organizations with a higher proportion of women in management tend to achieve improved financial performance, highlighting the positive impact of gender diversity on overall organizational success.

The positive correlation between the intention of women's behaviour towards technology in logistics and the respondent's overall intention aligns with the increasing presence of women in the logistics field. Authors of [3] emphasize that women actively pursue logistics careers and possess the necessary knowledge, skills, sensitivity, empathy, and qualifications. Women's contributions to creating a more balanced and comfortable environment challenge the traditional male-dominated landscape of the sector.

The finding that the perception of logistics culture significantly influences women's behaviour towards technology in logistics is consistent with the importance of work–family balance, as [9] suggested. Addressing work–family conflicts is vital for the well-being and success of both men and women working in logistics.

Finally, the positive influence of perception of women's characteristics and logistics culture on women's behaviour towards technology in logistics aligns with the research conducted by [10]. Workplace relationships can impact women's career success, and their unique relationship-oriented characteristics contribute to their effectiveness in supply chain management. Women's ability to see the bigger picture, communicate effectively, focus on goals and problem-solving, and their creativity, collaboration, and courage add substantial value to the logistics industry.

In conclusion, the findings derived from this study provide valuable insights into the factors influencing women's behaviour towards technology in the logistics sector. These findings are further supported by existing research, highlighting the importance of fostering a supportive and inclusive environment for women pursuing careers in logistics. Recognizing the significance of attitudes, knowledge, subjective norms, perceived behaviour control, and perception of logistics culture and women's characteristics is crucial for promoting women's active participation and success in the logistics industry (Table 1).

Table 1 Result of hypothesis verification

Hypothesis	Path	T statistics	P values	Decision
H1	Attitude → knowledge	4.924	0.000	Supported
H2	Intention → women behaviour	4.890	0.000	Supported
H3	Knowledge → intention	2.929	0.003	Supported
H4	Knowledge → voluntary	16.333	0.000	Supported
H5	Logistics culture → women behaviour	2.953	0.003	Supported
H6	Perceived behaviour control → intention	3.368	0.001	Supported
H7	Subjective norms → intention	2.100	0.036	Supported
H8	Voluntary → intention	4.426	0.000	Supported
H9	Women characteristic → logistics culture	9.977	0.000	Supported
H10	Women culture → women behaviour	5.921	0.000	Supported

4 Conclusion

This research study lays a comprehensive theoretical foundation for implementing women's responsibilities within the logistics industry. The researcher emphasizes actively involving the target population throughout the research process to ensure the study's relevance and applicability to their experiences. Furthermore, the study underscores the importance of examining the antecedent beliefs associated with the three constructs of the theory of planned behaviour (TPB). This critical analysis not only determines whether the TPB constructs account for variations in behavioural intentions but also provides insights into the underlying reasons behind women's engagement or lack thereof in specific environmental behaviours. This deeper understanding is instrumental in shedding light on the factors influencing women's choices and actions within the logistics industry. Importantly, the TPB framework can be effectively employed in any organizational context to explore how women can make the logistics sector more sustainable.

From a practical standpoint, the study's findings have yielded significant benefits for the logistics industry. The identified characteristics of women's behaviour serve as valuable guidance for developing targeted initiatives to improve gender inclusivity within the sector [11]. Moreover, the study's robust and reliable data collection methods enable organizations to accurately assess the impact of increased female representation in supply chain and logistics positions, potentially providing them with a competitive advantage. By incorporating the study's main conclusions into actionable processes, companies demonstrate their commitment to utilizing customized measures and leveraging the TPB framework to address barriers and facilitators of women's behaviour in the workplace. This integration fosters an environment where women can thrive and contribute to the broader objectives of the logistics industry.

The study also offers valuable recommendations for future research endeavours to understand the substantial underrepresentation of women in logistics professions. Exploring how workplace atmosphere and logistics stereotypes influence women's career aspirations and goals is essential. Investigating potential moderators, such as feminist identity development, on the relationship between perceived subjective norms and logistics career intentions can provide further insights into women's decision-making processes. Furthermore, future research should adopt a more qualitative approach, utilizing interviews, case studies, focus groups, and observations to complement the predominantly quantitative designs in existing studies. Expanding the geographical scope of research to include diverse locations and communities grappling with similar challenges would contribute to a more comprehensive understanding of women's role in driving technology adoption within the logistics sector. Lastly, future studies should stay abreast of the latest issues raised by relevant authorities, ensuring that research remains timely and aligned with the industry's evolving dynamics.

Acknowledgement We extend our sincerest gratitude to all the respondents who participated in this research.

References

1. R.V. Krejcie, D.W. Morgan, Determining sample size for research activities. Educ. Psychol. Meas. **30**, 607–610 (1970)
2. J.F. Hair, G.T.M. Hult, C.M. Ringle, M. Sarstedt, *A Primer on Partial Least Squares Structural Equation Modeling (PLS-SEM)*, 3rd edn. (Sage, Thousand Oaks, 2022)
3. S. Ruel, M.C.F. Morgane, Gender diversity in supply chains: towards more sustainable decisions? evidence from interviews. Supply Chain Forum. **22**(2) (2020). https://doi.org/10.1080/16258312.2021.1948307
4. A.C. Johnson, Women and minorities in transportation: Changing demographics and the impacts of Title VI and Title IX. J. Public Transpot. **19**(1), 81–96 (2016)
5. P. Sheeran, D. Trafimow, C.J. Armitage, Predicting behaviour from perceived behavioural control: Tests of the accuracy assumption of the theory of planned behaviour. British J. Soc. Psych. **41**(3), 393–410 (2002)
6. R. Aquilon, Competences in the logistics and transport sectors, in *Green Transportation Logistics*, (Springer, 2011), pp. 33–58
7. A.M. Knemeyer, P.R. Murphy, M.D. Pagell, Understanding the complexities of reverse logistics. Bus. Horizons. **42**(6), 61–71 (1999)
8. Y.M. Chin, H.H. Tat, Does gender diversity on corporate boards really boost firm performance? evidence from Malaysia. Corp. Own.Ship Control. **13**(1), 289–304 (2015)
9. M.J. Maloni, S.R. Sarder, U. Raja, The influence of supply chain work culture on women's work-family conflict and satisfaction. J. Bus. Logist. **40**(4), 341–355 (2019)
10. X. Gao, Y. Wu, Relationships among workplace friendship, gender, perceived career development, and career success. Int. J. Hum. Resour. Manage. **25**(15), 2143–2157 (2014)
11. N.K.A. Azlin, H.R.M. Sapry, J. Jaafar, J.M. Ghazali,A.R. Ahmad, Women's behavior toward the supply chain roles in Malaysia. Springer Briefs Appl. Sci. Technol. 69–74 (2023)

The Review and the Effect of Implementations: The Lean Six Sigma with Industrial Revolutions 4.0 in Logistics Industry

Hanafi Ahmad, Mohd Yazid Md Taib, Hairul Rizad Md Sapry, and Nurul Fadly Habidin

Abstract This paper's objective is to investigate the possibility of combining fourth industrial revolution (IR 4.0) concepts with lean six sigma approaches. While lean six sigma has long been seen as a successful system in the service industry, the advent of IR 4.0 has made logistics companies investigate the potential of lean six sigma approaches in more detail. IR 4.0 facilitates the integration of an intelligent factory, coordinating machines, human resources, products, and cloud-based solutions throughout the entire value chain. Within the logistics sector, notably in courier services like the POS logistics company, LSS principles and methodologies have been in use for many years. These practices have played a pivotal role in eliminating inefficiencies, cutting operational costs, and enhancing service performance. Services and manufacturers alike may reap double benefits by combining lean six sigma with industry 4.0 technology. First, this integration empowers services to boost productivity and enhance product quality by optimizing the implementation chain. Second, it facilitates self-management of operational processes, ultimately ensuring a higher level of product quality for customers.

Keywords Industrial revolutions 4.0 · Lean six sigma · Logistics performance

---

H. Ahmad (✉) · M. Y. M. Taib · H. R. M. Sapry
Industrial Logistics, Malaysian Institute of Industrial Technology, Universiti Kuala Lumpur, Masai, Johor, Malaysia
e-mail: hanafi@unikl.edu.my

M. Y. M. Taib
e-mail: mohdyazid@unikl.edu.my

H. R. M. Sapry
e-mail: hairulrizad@unikl.edu.my

H. Ahmad · N. F. Habidin
Faculty of Management and Economics, University Pendidikan Sultan Idris, Tanjung Malim, Perak, Malaysia
e-mail: fadly@upsi.edu.my

A. Ismail et al. (eds.), *Industrial Logistics I*,
SpringerBriefs in Applied Sciences and Technology,
https://doi.org/10.1007/978-3-032-26981-2_9

1 Introduction

The changing demands of consumers, who now want shorter turnaround times, higher product quality, and many other obstacles that supply chain and logistics companies must overcome pushing for greater economic innovation [1–4]. These organizations are always searching for methods to increase value and income while decreasing complexity and waste. They also evaluate the entire process of the supply chain to improve corporate performance [5]. Concurrently, the adoption of digital technologies has exerted a significant influence, and various advanced manufacturing paradigms have been developed to improve manufacturing processes and systems with a level of "intelligence" or "smartness," see [6–12]. Concurrently, organizations are harnessing digital technology within lean six sigma 4.0 optimization in real time, and data-driven decision-making is made possible by the work of [13, 14]. They have the potential to be a very formidable instrument when combined. Numerous studies have noted that there is great potential for generating improvements and boosting overall business performance through the confluence of industry 4.0 technology with lean six sigma methods [15–18]. This study uses a review analysis to determine the connections between lean six sigma and industry 4.0 logistics and supply chain organizations. This method is appropriate when a particular research issue has to be enhanced in order to avoid repeating studies on the same subject. The distinctive aspect of this research lies in its comprehensive exploration of the relationship between lean six sigma and industry 4.0, particularly in the context of various industries. Prior to this study, only a limited number of researchers had briefly examined this connection.

The study's outcomes are expected to contribute to the formulation of a robust framework for seamlessly integrating lean six sigma and industry 4.0 within enterprises characterized by diverse organizational cultures and processes. The central scientific proposition of this document underscores the necessity for the systematic development of knowledge regarding the fusion of lean six sigma and industry 4.0. This systematic development is deemed essential as it would pave the way for crafting a comprehensive conceptual framework for their joint implementation. Furthermore, this establishment of a foundational framework is expected to serve as a bedrock for guiding future research endeavors in this domain. The above statement has risen a research question on the combinations of lean six sigma with industry 4.0 on the logistics performance.

Research Questions: What is the effect of the combination of lean six sigma with IR 4.0 on the logistics performance?

This paper is organized as follows. Section 1 contains the introduction. In the Sect. 2 we have a brief description of industry 4.0 and its technologies, as well as a description of lean six sigma, and the effectiveness lean six sigma and industry 4.0. The results regarding the impact of lean six sigma with industry 4.0 technologies are discussed in Sect. 3. Section 4 concludes the document and provides future research directions.

2 Overview on IR 4.0 and Lean Six Sigma

2.1 Industry 4.0

Throughout history, emerging technologies have consistently spearheaded communication, real-time interactions, and flexibility, particularly since the advent of the industrial revolution. These technologies have played a pivotal role in driving progress, enabling notable achievements such as the reduction of costs and downtime. Innovation has already been used to describe industry 4.0 [19], and industry 4.0, as defined by [20], provides a new approach to rise the product quality and boosting process effectiveness, but industry 4.0 is referred to collectively by [21] as a set of technologies used by value chain companies. According to [22], this integration allows for data-driven decision-making and helps optimize production processes.

A notable achievement has been attained through the recent convergence of technological progress, culminating in the establishment of the industry 4.0 standard [23, 24] have identified a roster of 13 distinct industries 4.0 tools, augmenting the core nine technologies that define the landscape of industry 4.0. In addition, [20] delves into the importance of specific technologies including cloud computing, IoT, big data analytics, cyber-physical systems, and machine-to-machine communication. Indeed, these technologies are positioned to bolster organizations' capabilities in efficiently overseeing logistics and supply chain projects.

2.2 Lean Six Sigma

Presently, lean six sigma principles have been implemented across diverse industry sectors, spanning manufacturing, banking, health care, information technology (IT), government, and even non-profit organizations; lean six sigma approaches have become more popular in the last several decades [24], and overall supply chain performance assessment is one of the ways organizations have taken the initiative to boost logistics and supply chain performance [5, 21].

Organizations of all sizes are embracing lean six sigma as a driving force behind their digital transformation efforts [23], integration offers the benefit of integrating lean waste reduction strategies with six sigma variability reduction methods [25]. Nevertheless, there are certain limitations associated with lean six sigma that practitioners in the field grapple with, as indicated [22]. These assert that to effectively confront modern challenges, such as escalating complexity and vast volumes of data, companies must embrace digital technologies to their fullest extent to retain competitiveness [21].

2.3 *The Effectiveness of IR 4.0 and Lean Six Sigma on Supply Chains*

The essence of lean six sigma philosophy centers around crafting an efficient production system that minimizes waste, produces high-quality goods, and optimizes resource utilization [21]. The fourth industrial revolution also helps make production tools smarter and more efficient [20], which boosts efficiency and pride for customers. The improvement of company performance is the common goal of both lean six sigma and industry 4.0 [1]. Firms may unleash significant improvement potential by connecting I4.0 technology with lean and six sigma (SS) processes. This alignment helps companies achieve greater performance [15–17, 20]. The combination of industry 4.0's capabilities with continuous improvement approaches like lean six sigma has the potential to drive incremental improvements and accelerate enterprises to achieve outstanding performance.

Once again, when looked at independently, lean six sigma and Industry 4.0 are powerful tools for improving company performance and processes [26] extraordinarily potent and impactful tool [1]. For instance, many researchers have played an active role in this field's investigation and development, including scholars like [1, 15, 18, 20]. These technologies offer organizations enhanced capabilities for managing lean six sigma projects by improving time management and facilitating streamlined data accessibility [26]. Furthermore, they encourage the use of e-business technology and the exchange of information [20].

Hence, the integration of industry 4.0 technology with lean six sigma is expected to fortify the principles of continuous improvement. Consequently, it is reasonable to deduce that merging lean six sigma and Industry 4.0 strategies within an organization can complement each other's strengths [16, 18]. Notably, researchers have also delved into distinct obstacles encountered while implementing industry 4.0 [19]. Recognizing the impact of lean six sigma on the performance of logistics and supply chains is imperative, as it offers insights into the era marked by digitally optimized operational processes.

3 Results and Discussion

The obtained results provide a comprehensive overview of the literature that substantiates and aligns with these perspectives. The findings from [1] highlight the synergistic relationship between lean six sigma and industry 4.0. They suggest that companies with a robust LSS culture are better positioned to facilitate the transition to industry 4.0. Simultaneously, industry 4.0 technologies can offer enhanced performance benefits to companies that already employ lean six sigma methodologies. The combinations of these methodology allowed to better manage LSS projects in time and data accessibility [26]. Nevertheless, it is important to note that there exist additional technologies frequently linked with lean six sigma 4.0, as observed across various systematic review studies. These technologies, while not extensively explored within this section, have been recognized by multiple authors as having relevance to lean six sigma 4.0 methodologies.

The findings strongly indicate a natural synergy between the majority of lean six sigma and industry 4.0 technologies. This synergy not only aids in effective problem-solving but also fosters a culture of accountability, ultimately leading to enhancements in service delivery and reduction of wasteful human activities and resource utilization, while simultaneously elevating the quality of the customer experience and diminishing costs. This holistic approach to integration showcases its potential to revolutionize processes within various domains, promoting both sustainability and excellence.

4 Conclusion

The study delves into the intersection of lean six sigma and industry 4.0. Through this extensive dataset, the study aims to unravel the trajectory of development and intricate relationships within the LSS and Industry 4.0 landscape. This approach is well suited for uncovering and understanding the intricate relationships, and shared terminologies between these two domains. Furthermore, the competitiveness of a company can experience a substantial boost through the augmentation of production flexibility and efficiency. This enhancement is achieved by fostering seamless communication, information dissemination, and the infusion of intelligence into operations. This transformative process entails the interconnectedness of various components, leading organizations to establish intelligent networks across the value chain. These networks possess the capability to operate autonomously, culminating in a more streamlined, responsive, and efficient production ecosystem.

References

1. J. Antony, O. McDermott, M. Sony, Quality 4.0 conceptualization and theoretical understanding: A global exploratory qualitative study. TQM J. (2022). https://doi.org/10.1108/TQM-07-2021-0215
2. A. Cherrafi, S. Elfezazi, K. Govindan, J.A. Garza-Reyes, K. Benhida, A. Mokhlis, A framework for the integration of green and lean six sigma for superior sustainability performance. Int. J. Prod. Res. **55**(15), 4481–4451 (2017)
3. B.A. Lameijer, W. Pereira, J. Antony, The implementation of lean six sigma for operational excellence in digital emerging technology companies. J. Manuf. Technol. Manag. **32**(9), 260–284 (2021). https://doi.org/10.1108/JMTM-09-2020-0373
4. E. Psomas, J. Antony, Research gaps in lean manufacturing: A systematic literature review. Int. J. Qual. Reliab. Manag. **36**(5), 815–839 (2019)
5. Y.D. Putri, L.N. Huda, S. Sinulingga, The concept of supply chain management performance measurement with the supply chain operation reference model (journal review). IOP Conf. Ser.: Mater. Sci. Eng. **505**(1), 01201 (2019)

6. F. Tao, Q. Qi, L. Wang, A.Y.C. Nee, Digital twins and cyber–physical systems toward smart manufacturing and industry 4.0: Correlation and comparison. Engineering **5**(4), 653–661 (2019)
7. K. Thoben, S. Wiesner, T. Wuest, Industry 4.0 and smart manufacturing—A review of research issues and application examples. Int. J. Autom. Technol. **11**(1), 4–16 (2017)
8. Y. Liu, X. Xu, Industry 4.0 and cloud manufacturing: A comparative analysis. J. Manuf. Sci. Eng. **139**(3), 034701 (2017)
9. R.Y. Zhong, X. Xu, E. Klotz, S.T. Newman, Intelligent manufacturing in the context of industry 4.0: A review. Engineering **3**(5), 616–630 (2017)
10. Q. Qi, F. Tao, Digital twin and big data towards smart manufacturing and industry 4.0: 360-degree comparison. IEEE Access **6**, 3585–3593 (2018)
11. A. Malaga, S. Vinodh, Benchmarking smart manufacturing drivers using Grey TOPSIS and COPRAS-G approaches. BIJ **28**(10), 2916–2951 (2021)
12. S. Talapatra, M.K. Uddin, J. Antony, S. Gupta, E.A. Cudney, An empirical study to investigate the effects of critical factors on TQM implementation in the garment industry in Bangladesh. Int. J. Qual. Reliab. Manag. **37**(9–10), 1209–1232 (2019)
13. G. Heyes, M. Sharmina, J.M.F. Mendoza, A. Gallego-Schmid, A. Azapagic, Developing and implementing circular economy business models in service-oriented technology companies. J. Clean. Prod. **177**, 621–632 (2018). https://doi.org/10.1016/j.jclepro.2017.12.168
14. L. de Sousa, A.B. Jabbour, C.J.C. Jabbour, M. Godinho Filho, D. Roubaud, Industry 4.0 and the circular economy: A proposed research agenda and original roadmap for sustainable operations. Ann. Oper. Res. **270**(1–2), 273–286 (2018)
15. C. Anass, B. Amine, E.H. Ibtissam, I. Bouhaddou, S. Elfezazi, Industry 4.0 and lean six sigma: Results from a pilot study, in *International Conference on Integrated Design and Production*, (Springer, Cham, 2019), pp. 613–619
16. H.S. Sodhi, D. Singh, B.J. Singh, Effect of lean six sigma in small and medium enterprises of India: A review. Adalya J. **8**(8), 57–67 (2019)
17. S.H. Park, S.M. Dahlgaard-Park, D.C. Kim, New Paradigm of lean six sigma in the 4th industrial revolution era. Qual. Innov. Prosper. **24**(1), 1–16 (2020)
18. S. Tissir, A. Cherrafi, A. Chiarini, S. Elfezazi, S. Bag, Lean six sigma and industry 4.0 combination: Scoping review and perspectives. Total Qual. Manag. Bus. Excell. **34**(3–4), 1–30 (2022)
19. C.O. Klingenberg, M. Antonio, V. Borges, Industry 4.0 as a data-driven paradigm: A systematic literature review on technologies. J. Manuf. Technol.. Management, ahead-of-print, (2018). https://doi.org/10.1108/JMTM-09-2018-0325
20. G.L. Tortorella, D. Fettermann, Implementation of industry 4.0 and lean production in brazilian manufacturing companies. Int. J. Prod. Res. **56**(8), 2975–2987 (2018). https://doi.org/10.1080/00207543.2017.1391420
21. M. Hermann, T. Pentek, B. Otto, Design principles for Industrie 4.0 scenarios. Proceedings of the Annual Hawaii International Conference on System Sciences 2016 (2016), pp. 3928–3937
22. M. Núñez Merino, M. Maqueira, M. Juan, J. Moyano-Fuentes, P. Martínez-Jurado, Industry 4.0 and lean supply chain management, in *Enabling Technologies for the Successful Deployment of Industry 4.0*, ed. by A. Sartal, D. Carou, J.P. Davim, (CRC Press, Boca Raton, FL, 2020), pp. 105–128
23. C.O. Klingenberg, M.A.V. Borges, J.A.V. Antunes, Industry 4.0 as a data-driven paradigma systematic literature review on technologies. J. Manuf. Technol. Manag. **32**(3) (2019). https://doi.org/10.1108/JMTM-09-2018-0325
24. M. Rüßmann, M. Lorenz, P. Gerbert, M. Waldner, J. Justus, P. Engel, M. Harnisch, Industry 4.0: The future of productivity and growth in manufacturing industries. Germany—DE. https://www.bcg.com/de-de/publications/2015/engineered_products_project_business_industry_4_future_productivity_growth_manufacturing_industries (2015)
25. A.B. Singh, G. Gaurav, P.K. Sarkar, G.S. Dangayach, M.L. Meena, Present, past, and future of lean six sigma applications: from evolution to the era of artificial intelligence. Recent Pat. Eng. (2023)

26. B.N. Pasi, S.K. Mahajan, S.B. Rane, The current sustainability scenario of industry 4.0 enabling technologies in Indian manufacturing industries. Int. J. Product. Perform. Manag. **70**(5), 1017–1048 (2021). https://doi.org/10.1108/IJPPM-04-2020-0196

A Bibliometric Analysis: The Effect of Lean Six Sigma 4.0 on Logistics Performance

Hanafi Ahmad, Ahmad Fakhruddin Ahmad Dailami, Mohd Yazid Md Taib, Hairul Rizad Md Sapry, and Nurul Fadly Habidin

Abstract Businesses have put a lot of effort into incorporating lean six sigma (LSS) and industrial revolution 4.0 (IR 4.0) into their evaluations throughout the last decade. However, a recent investigator uncovered certain information concerning ground-breaking technologies that emphasize the requirement for defining and comprehending the role of the quality function in the quest for achieving excellence. This paper aims to investigate the future directions of LSS by conducting a bibliometric analysis study of LSS in the logistics services sector. It will also outline a research agenda for future LSS method research in these industries. To accomplish this objective, a grand total of 508 academic papers, spanning the period from 2005 to 2023, were collected from Scopus databases. These papers underwent thorough scrutiny employing a range of techniques, analysis of bibliometric indicators, cluster analysis, and network analysis. The logistics and supply chain subsector of the services industry was the primary focus of the investigation. Furthermore, the investigation identified the 15 most widely recognized journals relevant to the subject matter. The inquiry unveiled relevant publications, authors, and articles that have extensively addressed lean six sigma projects within services sectors. The study's findings will assist practitioners in understanding the specifics of these industries and identify

H. Ahmad (✉) · A. F. A. Dailami · M. Y. M. Taib · H. R. M. Sapry
Industrial Logistics, Malaysian Institute of Industrial Technology, Universiti Kuala Lumpur, Masai, Johor, Malaysia
e-mail: hanafi@unikl.edu.my

A. F. A. Dailami
e-mail: ahmad.fakhruddin@unikl.edu.my

M. Y. M. Taib
e-mail: mohdyazid@unikl.edu.my

H. R. M. Sapry
e-mail: hairulrizad@unikl.edu.my

N. F. Habidin
Faculty of Management and Economics, University Pendidikan Sultan Idris , Tanjung Malim, Perak, Malaysia
e-mail: fadly@upsi.edu.my

A. Ismail et al. (eds.), *Industrial Logistics I*,
SpringerBriefs in Applied Sciences and Technology,
https://doi.org/10.1007/978-3-032-26981-2_10

areas that need more research. The research agenda created by this study highlights pertinent subjects, techniques, and industries based on significant papers and writers, and the most used terms. Overall, this study provides valuable insights into the current state of LSS and future directions for research and implementation in the manufacturing and services industries.

Keywords Industrial revolution 4.0 · Lean six sigma · Logistics performance

1 Introduction

The changing demands of consumers for faster delivery times, higher quality products, and individualized service are just a few of the obstacles that supply chain and logistics companies are struggling to meet. Moreover, these businesses are also fighting increased levels of rivalry for larger portions of the market and promoting new forms of economic innovation [1, 2]. The main emphasis of innovative organizations revolves around achieving and maintaining competitiveness. These organizations are always searching for methods to increase value and income while decreasing complexity and waste. They also evaluate the entirety of the supply chain to improve corporate performance [3]. Concurrently, the adoption of digital technologies has exerted a significant influence on the implementation of process improvement methodologies over the past decade. Several advanced manufacturing paradigms have emerged, drawing from [4–6] with the goal of imbuing manufacturing processes and systems with an element of "intelligence" or "smartness." At the same time, businesses are utilizing digital technology in closed-loop supply chains to address both the "end-of-life" concept in the industrial system and the need for restoration and regeneration. Implementing lean six sigma allows for real-time optimization of the production process and data-driven decision-making [7–11]. Numerous researchers have noted that there is great potential for generating improvements and boosting overall business performance through the confluence of industry 4.0 technology with lean six sigma methods.

The study's outcomes are intended to contribute to the formulation of a robust framework for seamlessly integrating of LSS and IR 4.0 within enterprises characterized by diverse organizational cultures and processes. Furthermore, this research is anticipated to foster waste reduction, encompassing not only traditional forms of waste but also digital waste. The central scientific proposition of this document underscores the necessity for the systematic development of knowledge regarding the fusion of LSS and IR 4.0. This systematic development is deemed essential as it would pave the way for crafting a comprehensive conceptual framework for their joint implementation. Furthermore, this establishment of a foundational framework is poised to serve as a bedrock for guiding future research endeavors in this domain.

2 Overview on IR 4.0 and Lean Six Sigma

2.1 *Industrial Revolutions*

Throughout history, emerging technologies have consistently spearheaded communication, real-time interactions, and flexibility, particularly since the advent of the industrial revolution. These technologies have played a pivotal role in driving progress, enabling notable achievements such as the reduction of costs and downtime. Innovation has already been used to describe IR 4.0 [11]. However, [9] defines it as providing a new approach to raising product quality and boosting process effectiveness, but IR 4.0 is referred to collectively by as a set of technologies used by value chain companies. Cyber-physical systems (CPS), the internet of things (IoT), the internet of services (IoS), and the smart factory are its basic components. Furthermore, IR 4.0 facilitates the seamless integration of information across the logistics landscape, characterized by real-time processing. According to the research conducted by [13], this integration enables decision-making based on data and helps optimize production processes. Authors of [14] have identified a roster of 13 distinct industries 4.0 tools, augmenting the core nine technologies that define the landscape of industry 4.0. System simulation, horizontal and vertical system integration, autonomous robots, cloud computing, big data analysis, cyber-physical systems, additive manufacturing (3D printing), internet of services (IoS), augmented reality (AR), internet of people (IoP), blockchain, and internet of data (IoD) are all part of these toolsets [15]. In fact, these technologies are poised to enhance businesses' capacities to successfully manage logistics and supply chain.

2.2 *Lean Six Sigma*

Lean six sigma has become more popular in the last several decades as a strategy for companies looking to succeed in today's demanding market [16]. Organizations are taking the initiative [17] to improve logistics and supply chain performance in many ways. The applications of LSS in logistics and supply chain have gained attention on the application of the LSS methodology. Organizations of all sizes are embracing LSS as a driving force behind their digital transformation efforts [18]. By bringing together lean waste reduction and six sigma variation reduction, integration offers several benefits, as noted [19]. However, as said, there are a few limitations of lean six sigma that professionals in the industry must deal with [20] complexity and massive amounts of data [21]. The capability of LSS has not yet measure to gain some benefit and resolve among the challenge face.

2.3 *The Effectiveness IR 4.0 and Lean Six Sigma on Logistic*

The lean six sigma philosophy is to design a supply chain that maximizes resource usage, reduces waste, and creates high-quality commodities [22]. Similarly, IR 4.0 improves operational performance and customer satisfaction by enabling the development of production tools into intelligent and efficient entities [21]. In other words, improving company performance is the common goal of LSS and IR4.0 [1]. This alignment aids in attaining greater performance [21]. By combining industry 4.0's potential with continuous improvement approaches like lean six sigma, businesses may achieve outstanding levels of performance.

Improvements in supply chain performance may be driven by IR 4.0 and its supporting technologies, according to [14]. By streamlining data accessibility and enhancing time management, these technologies provide businesses with better capabilities for managing LSS initiatives [15], and they also foster information sharing [10] the utilization of e-business technologies [18]. The performance of supply chain has related to the integrations on the methodology and technology. This leads us to believe that IR 4.0 and lean six sigma, when implemented together, may be a powerful asset to any company [20]. The implementation of IR 4.0 has not been without its own challenges, which scholars have thoroughly examined [21]. Since IR 4.0 heralds the age of digitally optimized operational processes, it is critical to acknowledge its influence on logistics and supply chain performance.

The following is the outline of the paper. You may find the introduction in Sect. 1. Section 2 provides an overview of industry 4.0 and its technologies, while Sect. 3 details the approach employed, which is systematic review analysis. Sections 4 and 5 highlight the outcomes of integrating LSS with industry 4.0 technology. Section 5 offers suggestions for further study and serves as the document's conclusion.

3 Methodology

The objective of our study is to find papers that explain how lean six sigma approaches are used with industry 4.0 technology. Examining these publications will help us validate and understand how LSS and industry 4.0 technologies work together. Most of these procedures are involved in a systematic literature review: By creating a review process that is systematic, transparent, and repeatable, bibliometric analysis has the potential to increase the quality of reviews. This article provides an in-depth exploration of both primary and secondary research steps. The research methodology involves a comprehensive investigation into the terms "lean six sigma" and "industry 4.0." To effectively gather relevant information, a specific query was devised for Scopus, a database known for its extensive collection of cited references. This choice was made due to Scopus's capacity to offer comprehensive information on cited sources.

4 Results and Discussion

4.1 Research Growth: Evolutions of Independency Between the IR 4.0 and Lean Six Sigma

There is a clear pattern to the annual citation counts of these works. Between 2015 and 2022, the number of published papers on lean six sigma and industry 4.0 increased consistently. During this time span, the number of studies typically ranges from 20 to 70 every year, with 2021 recording the greatest number of research at 55.

4.2 Subject Area of IR 4.0 with the Lean Six Sigma

Lean six sigma and industry 4.0 related papers were compiled for this study from Scopus and the Web of Science databases. The main objective of this study was to identify the most common areas of research into LSS and IR 4.0. The results of this investigation demonstrate how widely used these concepts are in many different fields of study.

4.3 Geographical Distributions of IR 4.0 and Lean Six Sigma Publications

Significant regional differences exist in the distribution of publications pertaining to lean six sigma and industry 4.0, which is worth mentioning. Germany has 44 papers, which is quite a number, followed closely by Italy with 22 papers. After this comes the contributions from China, with 32 papers, Brazil with 16, and the USA with 15. Surprisingly, Japan does not seem to be making any contributions to the productivity of this study subject.

4.4 Bibliographic Network Analysis

4.4.1 Citations Network Analysis

Scopas and WOS reports that to determine which linkages and research subjects were referenced in the 300 papers under consideration, a network citation analysis was carried out. As part of the network citation analysis, the writers used the VOS viewer to map the keywords supplied by each article.

4.4.2 Keywords Analysis

The researchers visually mapped the given keywords for each unique article using the VOS viewer programmed for the purpose of keyword analysis. The keywords, visually represented as distinct-colored spheres of varying sizes, are traditionally showcased individually. However, their inherent interconnectedness becomes evident upon closer examination. This synthesis highlights the intricate relationships among terms like "Industry 4.0," "LSS Manufacturing," "Factory of the Future," "Performance," and "Sustainability." Additionally, it underscores the correlation between specific Industry 4.0 and LSS technologies, such as "Big Data," "Logistics," "Supply Chain," and "Quality Management." These connections arise not only due to shared objectives and tools but also due to the association of "Logistics" with both LSS and industry 4.0, alongside other relevant keywords.

4.5 The Impact Between LSS 4.0 and Logistics Performance

Building upon the results obtained from the keyword analysis, we embarked on an exploration of the correlations between LSS tools and industry 4.0 technologies. We used the complex network analysis (CNA) as a framework to help with this inquiry. This strategic approach enabled us to gain a more profound comprehension of the intricate interplay, influences, and interconnections that exist between LSS methodologies, and the advancements associated with industry 4.0.

Due to the variation in the literature review, not all authors have addressed the combination of LSS 4.0 tools and logistics performance in their papers. As a result, a decision was made not to incorporate these aspects within the present article, despite having referenced their works during the writing process. It should be noted that some authors have covered all the features of LSS 4.0, while others have concentrated on only a few of the features. In addition, 21 of the total writers analyzed have written about how to use industry 4.0 tools in conjunction with lean six sigma. The evaluated literature places different amounts of emphasis on various characteristics, as seen by the variation in coverage. The growing importance of aligning LSS with industry 4.0 concepts to improve operational efficiency and overall performance is emphasized by the joint research efforts highlighted in this body of work.

The findings strongly indicate a natural synergy between the majority of LSS and industry 4.0 technologies. A culture of responsibility is fostered, and problem-solving is made more effective through this synergy, which in turn improves service delivery, decreases inefficient human activities and resource consumption, increases the quality of the patient experience, and decreases costs. This holistic approach to integration showcases its potential to revolutionize processes within various domains, promoting both sustainability and excellence.

5 Conclusion

This research investigates how LSS and IR 4.0 intersect. The research methodology hinges on the application of a bibliometric technique, which offers a systematic means to comprehend the evolution and underlying structures within this field. The primary data source for this study is Scopus and WOS as a robust academic database. The analysis is based on a collection of 508 academic papers, encompassing the timeframe from 2005 to 2023. Through this extensive dataset, the study aims to unravel the trajectory of development and intricate relationships within the LSS and IR 4.0 landscape. In fact, the bibliometrically-based technique is well known for its success in allowing a unified co-word analysis of the "Lean Six Sigma" and "Industry 4.0" convergence. The common vocabularies, complex webs of links, and overarching themes connecting these two fields may be better understood using this method. By utilizing this methodology, the study aims to enhance comprehension and insights into the interplay between lean six sigma and industry 4.0, thereby contributing to the broader understanding of their integration especially quantitative or qualitative research on the applications of LSS 4.0. This transformative process entails the interconnectedness of various components, leading organizations to establish intelligent networks across the value chain. These networks possess the capability to operate autonomously, culminating in a more streamlined, responsive, and efficient production ecosystem.

References

1. J. Antony, O. McDermott, D. Powell, M. Sony, The evolution and future of lean six sigma 4.0. TQM J. **35**(4), 1030–1047 (2023). https://doi.org/10.1108/TQM-04-2022-0135
2. E. Psomas, J. Antony, Research gaps in lean manufacturing: A systematic literature review. Int. J. Qual. Reliab. Manage. **36**(5), 815–839 (2019). https://doi.org/10.1108/IJQRM-12-2017-0260
3. Y.D. Putri, L.N. Huda, S. Sinulingga, The concept of supply chain management performance measurement with the supply chain operation reference model (journal review). IOP Conf. Ser.: Mater. Sci. Eng. **505**(1), 012011 (2019)
4. F. Tao, Q. Qi, L. Wang, A.Y.C. Nee, Digital twins and cyber–physical systems toward smart manufacturing and industry 4.0: Correlation and comparison. Engineering **5**(4), 653–661 (2019)
5. A. Malaga, S. Vinodh, Benchmarking smart manufacturing drivers using Grey TOPSIS and COPRAS-G approaches. BIJ **28**(10), 2916–2951 (2021)
6. S. Talapatra, A. Gaine, Putting green lean six sigma framework into practice in a jute industry of Bangladesh: A case study. Am. J. Ind. Bus. Manag. **9**(12), 2168–2189 (2019)
7. G. Heyes, M. Sharmina, J.M.F. Mendoza, A. Gallego-Schmid, A. Azapagic, Developing and implementing circular economy business models in service-oriented technology companies. J. Clean. Prod. **177**, 621–632 (2018). https://doi.org/10.1016/j.jclepro.2017.12.168
8. L. de Sousa, A.B. Jabbour, C.J.C. Jabbour, M. Godinho Filho, D. Roubaud, Industry 4.0 and the circular economy: A proposed research agenda and original roadmap for sustainable operations. Ann. Oper. Res. **270**(1–2), 273–286 (2018). https://doi.org/10.1007/s10479-018-2772-8

9. C. Anass, B. Amine, E.H. Ibtissam, I. Bouhaddou, S. Elfezazi, Industry 4.0 and lean six sigma: Results from a pilot study, in *International Conference on Integrated Design and Production*, (Springer, Cham, 2019), pp. 613–619
10. S. Tissir, A. Cherrafi, A. Chiarini, S. Elfezazi, S. Bag, Lean six sigma and industry 4.0 combination: Scoping review and perspectives. Total. Qual. Manag. Bus. Excel. 1–30 (2022). https://doi.org/10.1080/14783363.2022.2043740
11. J. Antony, O. McDermott, M. Sony, Quality 4.0 conceptualization and theoretical understanding: A global exploratory qualitative study. TQM J. (2021). https://doi.org/10.1108/TQM-07-2021-0215
12. C.O. Klingenberg, M.A.V. Borges, J.A.V. Antunes, Industry 4.0 as a data-driven paradigm systematic literature review on technologies. J. Manuf. Technol. Manag. **32**(3) (2019). https://doi.org/10.1108/JMTM-09-2018-0325
13. M. Núñez-Merino, J.M. Maqueira-Marín, J. Moyano-Fuentes, P.J. Martínez-Jurado, Information and digital technologies of industry 4.0 and lean supply chain management: A systematic literature review. Int. J. Prod. Res. **58**(16), 5034–5061 (2020)
14. E. Hofmann, M. Rüsch, Industry 4.0 and the current status as well as future prospects on logistics. Comput. Ind. **89**, 23–34 (2017)
15. B.N. Pasi, S.K. Mahajan, S.B. Rane, The cur- rent sustainability scenario of industry 4.0 enabling technologies in Indian manufacturing industries. Int. J. Product. Perform. Manag. **70**(5), 1017–1048 (2021). https://doi.org/10.1108/IJPPM-04-2020-0196
16. G. Yadav, D. Seth, T.N. Desai, Prioritizing solutions for lean six sigma adoption barriers through fuzzy AHP-modified TOPSIS framework. Int. J. Lean Six Sigma **9**(3), 270–300 (2018)
17. J. George, V.M. Pillai, A study of factors affecting supply chain performance. J. Phys. Conf. Ser. **1355**(1), 012018 (2019). https://doi.org/10.1088/1742-6596/1355/1/012018
18. M. Singh, R. Rathi, J.A. Garza-Reyes, Analysis and prioritization of lean six sigma enablers with environmental facets using best worst method: A case of Indian MSMEs. J. Clean. Prod. **279**, 123592 (2021)
19. G. Onofrei, B. Fynes, H. Nguyen, A.H. Azadnia, Quality and lean practices synergies: A swift even flow perspective. Int. J. Qual. Reliab. Manag. **38**(1), 98–115 (2021). https://doi.org/10.1108/IJQRM-11-2019-0360
20. M. Sony, S. Naik, K.K. Therisa, Why do organisations discontinue lean six sigma initiatives?. Int. J. Qual. Reliab. Manag. **36**(3), 420–436 (2019). https://doi.org/10.1108/IJQRM-03-2018-0066
21. A. Sanders, C. Elangeswaran, J. Wulfsberg, Industry 4.0 implies lean manufacturing: Research activities in industry 4.0 function as enablers for lean manufacturing. J. Ind. Eng. Manag. **9**(3), 811–833 (2016)
22. A. Kuiper, R.H. Lee, V.J. Ham, R.J. Does, A reconsideration of Lean Six Sigma in healthcare after the COVID-19 crisis. Int. J. Lean Six Sigma (2021)

Factors Influencing the Demand for Commuter Service of Keretapi Tanah Melayu

Jamilahtun Md Ghazali, Haziq Alif Mohd Khairul Amin, and Hairul Rizad Sapry

Abstract Public transport is the economic backbone of countries. This is a result of the community's reliance on public transport to meet needs such as working and obtaining daily necessities. Malaysia is a developing country that provides services for public transport including buses, trains, and ferries for people to move from one place to another. The objective of this paper is to analyze the factors influencing demand for the KTM Commuter, which is a type of rail-based public transportation. This is due to a reduction in passenger demand for Malaysia's KTM commuter service. The area of study focused by the researcher is the commuter line route from Tanjung Malim station to KL Sentral. A quantitative approach is used in this study. A questionnaire was distributed via face to face, online platforms, and email to respondent. Results of this study show that the convenience factor came out as a factors affecting KTM commuter service out of the three factors studied. Based on the findings, the recommendation for the future was documented to serve as initial input improvement of KTM commuter demand.

Keywords KTM commuter service demand · Public transport · Passenger

1 Introduction

Public transport is still the preferred method for the society in Malaysia, when it comes to daily activities that directly affect economic activity such as work and meeting basic needs [1]. This is particularly relevant for commuting between one's

J. M. Ghazali (✉) · H. A. M. K. Amin · H. R. Sapry
Industrial Logistics, Malaysian Institute of Industrial Technology, Universiti Kuala Lumpur, Masai, Johor, Malaysia
e-mail: jamilahtun@unikl.edu.my

H. A. M. K. Amin
e-mail: haziq.khairul@s.unikl.edu.my

H. R. Sapry
e-mail: hairulrizad@unikl.edu.my

A. Ismail et al. (eds.), *Industrial Logistics I*,
SpringerBriefs in Applied Sciences and Technology,
https://doi.org/10.1007/978-3-032-26981-2_11

place of residence and their place of employment [2]. If the people use it effectively, it is also viewed as an alternate method to reduce congestion. The goal is to ensure that a public transport infrastructure is created such that each population in every city can employ mass transit easily [3]. To encourage community growth and facilitate the mobility of urban and suburban life, public transit is therefore important in the city area.

Instead of having a private automobile, using public transport may be the most affordable approach to get through financial hurdles [4]. The government provides a variety of public transport options to the general population, including ferries, ships, trains, taxis, buses, and metros [5]. However, the Keretapi Tanah Melayu (KTM) Commuter, Kuala Lumpur Monorail, Light-Rail Transit (LRT), and Mass Rapid Transit (MRT) together make rail one of the most effective ways for urban and suburban society to get around. The importance of public transit is continually emphasized as a key area that needs to be developed, examined, and monitored by all parties concerned to provide high-quality services.

To persuade people to choose a service, it is important to underline factors like quality, comfort, a fair price, and response speed [6]. This attraction in public transport services, specifically for commuters, needs to be addressed to understand the reasons that influence people to use KTM Commuter.

The aim of this study is to investigate the variables that affect riders' demand preferences for using the KTM commuter service. The research objectives are stated as following:

1. To identify the level of demand for KTM commuter services.
2. To evaluate the impact of the demand factor in relation to individual groups of passenger.

Three hypotheses were developed to facilitate the investigation in research objective ii as follows.

H1: There is a significant relationship between accessibility and passenger demand.

H2: There is a significant relationship between connectivity and passenger demand.

H3: There is a significant relationship between safety and passenger demand.

2 Methodology

A total of 98 respondents who participated in this study came from various backgrounds. The respondents were selected from those that are KTM commuter passengers. All respondents of the survey are the passengers that used KTM commuter to commute from one place to another for working and leisure. The data obtained from the survey were analyzed using partial least square, SmartPLS 3.0 software, which is deemed suitable for this study. The study has adopted three types of data analysis that include descriptive analysis, measurement model analysis, and structural model analysis to answer the research questions.

3 Results and Discussion

The role of researchers in examining data gathered from respondents is discussed in this section. Researchers decided to utilize descriptive analysis and the analysis tool known as partial least squares structural equation modeling (PLS-SEM) as measuring methodologies. Out of a sample size of 100 respondents, the researcher received data from 98 of them, which indicates a good response rate.

3.1 *Research Objective 1*

This objective research's aim is to investigate the level of passenger demand for KTM commuter service between Tanjung Malim station to KL Sentral. The goal also seeks to impact passenger demand through accessibility, connectivity, and safety. The respondent's data has been descriptively to get the mean and standard deviation values as per Table 1.

Respondents' levels of demand were generally modest, with mean values ranging from 2.00 to 2.02. There seems to be a variety of points of view based on the responses, which varied from 1 to 5. The median score of 2.00 indicates that most respondents were willing to pay more given a constrained range of pricing. The standard deviation, which ranged from 0.969 to 1.010, demonstrated quite little response dispersion. According to the research, consumers are generally willing to pay extra for enhancements in convenience, reliability, connectivity, hygiene, and safety.

Table 1 Passenger demand on KTM commuter services

Input	Mean	Minimum	Maximum	Median	Standard deviation
Would you be willing to pay more for this commuter service if it was in a more convenient location? (PSD1)	2	1	5	2	0.969
I am willing to pay more for train ticket if it means more frequent service and better connectivity. (PSD2)	2.01	1	5	2	1.01
I am willing to pay more for train tickets if it means a cleaner and more safety. (PSD3)	2.02	1	4	2	0.969

In Table 2, the information presented includes answers to questions about a terminal's accessibility. The responses are typically somewhat content or in agreement, according to the median values, which vary from 1.765 to 2.612. The minimum and maximum values, which vary from 1 to 5, show a spectrum of opinions. The median answers to most of the questions, which is 2.00, indicate that the answers are scattered quite evenly. There is a fair amount of response variability, as evidenced by the standard deviations, which vary from 0.831 to 1.007. Overall, the data shows that opinions on the terminal's comfort, ease of access, sufficiency of the facilities, and connection quality vary.

As per Table 3, the range of the mean values, from 1.622 to 2.765, reveals that the respondents' levels of satisfaction were variable. The minimum and highest values' range of 1–5 illustrates the great variety of opinions. The median results of 3.00 for questions about the number of trains during peak and off-peak hours represent a highly equally distributed response set. The range of standard deviations, from 0.851 to 1.219, shows a considerable amount of response variability. In general, the findings show discontent, particularly in relation to waiting times and the availability of trains outside of rush hours.

Table 2 Accessibility to KTM commuter service

Input	Mean	Minimum	Maximum	Median	Standard deviation
As a passenger, do you find the current terminal location convenient for accessing your destination place? (ACC1)	2.194	1	5	2	1.007
Do you think that the current terminal location makes it difficult to access certain parts of the city or surrounding areas? (ACC2)	2.3.27	1	4	2	0.945
Do you think the facilities at the terminal are sufficient for a comfortable travel experience? (ACC3)	2.429	1	5	2	0.892
Do you agree that the important is it for the terminal to have connections to different parts of the city or region via public transportation? (ACC5)	1.765	1	5	2	0.831
Do you think that the terminal provides convenient connections to other forms of public transportation? (ACC6)	2.194	1	5	2	0.853

Table 3 Convenience using KTM commuter service

Input	Mean	Minimum	Maximum	Median	Standard deviation
I believe that the frequency of train service is sufficient for my needs. (CON1)	2.694	1	5	3	1.155
I am satisfied with the number of trains offered during peak hours, but not during off-peak hours. (CON2)	2.765	1	5	3	1.219
I find it frustrating when I must wait a long time for the next train to arrive. (CO3)	1.622	1	5	3	0.851

Lastly, the data also includes answers to queries concerning safety in train stations in Table 4. The range of the mean ratings, from 1.602 to 2.367, reveals that there were differences in the respondents' levels of satisfaction or agreement. The minimum and highest values' range of 1–5 illustrates the great variety of opinions. The median values of 2.00 for most questions reflect an evenly distributed distribution of responses. The range of standard deviations, from 0.734 to 0.941, shows a considerable amount of response variability. Overall, the data shows that while some respondents express pleasure with safety procedures, cleanliness, and a sense of safety and security in railway terminals, others have concerns about signs, maintenance, and cleanliness.

3.2 *Research Objective 2*

Overall, for research objective 2, the analysis found significant connections between variables and high loading values, demonstrating their importance in influencing passenger demand for KTM commuter service. Convenience had the highest loading value, showing that it has a big impact on passenger demand. Strong correlations between the attributes of passenger demand, accessibility, and safety were also observed. High AVE and CR values demonstrate the model's dependability by demonstrating that the variables accurately reflect the components they are measuring.

The data show that hypothesis H1, which postulates a relationship between accessibility and passenger demand, is not supported by any evidence. The estimated T-value of 0.407 and corresponding p-value of 0.684 indicate that there is insufficient data to draw a firm conclusion regarding a relationship between accessibility and passenger demand. For the association between connectivity and passenger demand proposed in hypothesis H2 is supported. The moderate level of significance shown by the calculated T-value of 1.812 and p-value of 0.070 suggests that there is support

Table 4 Safety using KTM commuter service

Input	Mean	Minimum	Maximum	Median	Standard deviation
I feel safe and secure when using train terminals. (SAF1)	2.276	1	5	2	0.855
I am satisfied with the level of safety and security measures in place at train terminals, such as CCTV cameras and security personnel. (SAF2)	2.194	1	5	2	0.9
I believe that train terminals should provide clear and informative signage and announcements to promote safety and security awareness among passengers. (SAF3)	1.633	1	4	1	0.734
I believe that the train fleet is generally clean and well-maintained. (SAF4)	2.367	1	5	2	0.941
I find it unpleasant when the train environment is dirty or unkempt. (SAF5)	1.602	1	4	1	0.752

for the association between connectivity and passenger demand. H3, which seeks to prove a connection between passenger demand and safety, is unsupported. —The p-value of 0.692 and T-value of 0.396 proof no significant evidence for a relationship between safety and passenger demand, refer to Tables 5 and 6.

Table 5 Construct validity, dimensionality, reliability, and item

Variable	Item	Loading	VIF	AVE	CR	CA	R2
Passenger demand on KTM commuter service	PSD1	0.734	1.343	0.579	0.805	0.654	0.114
	PSD2	0.809	1.298				
	PSD3	0.738	1.187				
Accessibility	ACC1	0.739	1.131	0.572	0.801	0.628	
	ACC3	0.757	1.369				
	ACC4	0.774	1.370				
Convenience	CO1	0.927	1.711	0.821	0.901	0.810	
	C02	0.884	1.711				
Safety	SAF1	0.846	2.251	0.738	0.894	0.847	
	SAF2	0.879	2.563				
	SAF4	0.852	1.563				

Table 6 Hypotheses verification

Hypothesis	Relationship	T-Value	P Value	Decision
H1	ACC → PD	0.407	0.684	Not supported
H2	CON → PD	1.812	0.07	Supported
H3	SAF → PD	0.396	0.692	Not supported

4 Conclusion

The study's findings suggest that convenience is the primary driver of passenger demand for the KTM commuter service. This component has to do with problems like lengthy waits, delays, and frequent trains, especially during rush hour. Due to these issues, travelers from Tanjung Malim station to KL Sentral are disappointed and prefer to take private transportation. The goal of this study is to increase the number of operators of public transport, particularly those in the railway industry, to improve their customer services. Other researchers can use this study as a guide to find various aspects of various routes. This could raise the standard of the nation's public transport and increase demand for utilizing it.

References

1. A.M. Rohayu, S. Rosli, A. Jamalunlaili, N. Rohana, Q.J. Kwong, Factors influencing ridership preference in using LRT service in Klang Valley, Malaysia. J. Malays. Inst. Plan. **19**(2), 250–262 (2021)
2. J.P. Rodrigue, *The Geography of Transport Systems* (Routledge, London, 2020)
3. M. Buchanan, The benefits of public transport. Nat. Phys. **15**, 876–876 (2019)
4. I. Daud, Pengangkutan Awam Alternatif Jimat Belanja. Utusan Malaysia (2021). https://www.utusan.com.my/commentary/pengangkutan-awam-alternatif-jimat-belanja/(2021). Accessed 15 June 2023
5. Ministry of Transport Malaysia Infrastructure: Official Site Ministry of Transport Malaysia Infrastructure (2022). https://www.mot.gov.my/en/browse-by-topic/infrastructure. Access 15 June 2023
6. R. Muhamad Ali, Factors of choosing the use of KTMB commuters among workers at Kuala Lumpur. Int. J. Sci. Eng. Res. **12**(10), 211–2219 (2021)

Service Quality, Knowledge for Customer, and Customer Satisfaction Roles in Courier Services in Johor Bahru

Khairul Firdaus Adrutdin, Sallaudin Hassan, and Kamisan Gadar

Abstract This inquiry investigates key factors influencing customer satisfaction (CS) in courier services, focusing on service quality (SQ) and knowledge for customer (KC) in the Johor Bahru district. Utilizing a self-administered questionnaire, 384 respondents were involved, and partial least squares structural equation modeling (PLS-SEM) was engaged in analyzing the data. The research explores dimensions within SQ and KC, aiming to comprehend their impact on overall customer satisfaction. Results reveal significant associations between KC and SQ dimensions, highlighting their influence on customer satisfaction. This study enhances our understanding of courier service dynamics, underscoring the pivotal role of imparting knowledge to customers for satisfaction improvement. It offers practical insights for service providers seeking to optimize customer satisfaction through strategic enhancements in service quality and knowledge dissemination.

Keywords Service quality · Knowledge for customer · Customer satisfaction · Courier services

1 Introduction

With the courier services industry accounting for RM135.7 billion or approximately 8%, from the entire market capitalization of Bursa Malaysia, the industry has grown to be an important provider to the nation's economic wellbeing [1]. This growth is fueled in part by the surge in e-commerce and Internet-based businesses, a result

K. F. Adrutdin (✉) · S. Hassan
Malaysian Institute of Industrial Technology, Universiti Kuala Lumpur, Masai, Johor, Malaysia
e-mail: khairulfirdaus@unikl.edu.my

S. Hassan
e-mail: sallaudin@unikl.edu.my

K. Gadar
Universiti Kuala Lumpur Business School, Kuala Lumpur, Malaysia
e-mail: kamisan@unikl.edu.my

A. Ismail et al. (eds.), *Industrial Logistics I*,
SpringerBriefs in Applied Sciences and Technology,
https://doi.org/10.1007/978-3-032-26981-2_12

of technological advancements and shifts in consumer behavior influenced by the Covid-19 pandemic. Malaysians increasingly rely on online transactions, especially for daily essentials, utilizing courier services for deliveries. The convenience offered by this trend allows consumers to shop at their convenience and have purchases delivered to a location of their choice [2].

The courier services sector in Malaysia has witnessed notable advancements, sparking concerns regarding customer satisfaction and the need to align with customer expectations [3]. To ensure sustainable services, delivering high-quality products and services is paramount. Focusing on service quality improvement initiatives is vital, with a particular emphasis on timely package delivery [4]. Elevating satisfaction levels requires the incorporation of a customer-centric knowledge component, providing customers with insights into operational aspects. Customer knowledge plays a pivotal role in mediating interactions between service quality and satisfaction, preventing issues like prolonged delivery times and procedural inefficiencies. Besides fostering customer loyalty, the enhanced service quality can also influences purchasing decisions and the overall business performance of courier companies. While the conventional measurement of elementary service quality dimensions includes tangibility, empathy, assurance, responsiveness, and reliability, modifications are appropriate for more accurate measurement. To facilitate comparison, the traditional five-dimensional service quality model has been streamlined into three components.

2 Background of Studies

The inception of courier services in Malaysia traces back to the early 1800s when the government established a company through Pos Malaysia Bhd [5]. Since its founding in the early twentieth century, the nation's postal services have gradually increased in capacity and reach, absorbing private local and international courier services. Dominating a significant market share are well-established courier services like Pos Laju, J&T Express, Ninjavan, DHL, Citilink Express, among others. The Covid-19 pandemic has catalyzed an increased request for courier offerings, leading to increased competition within the industry. Pos Laju, for instance, has effectively delivered around 522 million postal items in recent years. According to a Rakuten Insight survey on pandemic-induced shopping behavior in Malaysia, 71 percent of respondents expressed specific time preferences for online shopping. Notably, J&T Express Malaysia has garnered recognition, securing two awards–"The Best Delivery Excellence Award" and "The Best Customer Service Excellence" awards from the Malaysian Communications and Multimedia Commission [3]. For a historical overview, refer to Table 1 for the timeline of courier business development in the country.

The significance of courier services for customers extends across various dimensions, including timely delivery, ease of use for customers, monitoring and openness, efficient wrapping, and economy of scale [5]. Conversely, within the framework of

Table 1 Malaysian courier services progress

Courier services provider	Established
Pos Malaysia Bhd	1868
DHL Malaysia	1974
City-link	1978
Nationwide	1986
ABX	1987
GDEX	1993
Ninja Van Malaysia	2015
Sky net	2016
J&T	2017

Malaysia, the significance of delivery services is extensive, encompassing the growth of e-commerce, expansion of businesses, revenue generation, facilitation of global trade, sustainability of supply chains, financial ramifications, creation of jobs, and the development of logistics facilities. This significance has led to the establishment of inter-company networks, combining of business activities, and improvements in the knowledge transfer between service providers and clients. Furthermore, courier services contribute to expenditures and confidence of consumers, offering services with additional value to the overall benefit of the market.

Courier services encompass the handling and management of the delivery of documents, parcels, and other items from one location to another, ensuring timely and secure delivery [6]. In response to technological advancements and evolving customer preferences, courier services have adopted cutting-edge technology to expand and diversify their service offerings. With applications like scanning barcodes and signature verification, consumers may track the delivery procedures in real time through courier services' incorporation of the newest technology, which speeds up the delivery process. Utilizing advanced mobile applications, customers can consistently monitor the progress of parcel or document deliveries at each stage of the process. This technological scheme devises the prospective in enhancing customer assertion and satisfaction with courier services, ultimately refining the overall quality of customer service [6].

Even though courier services have expanded quickly in Malaysia, the effects of globalization have made it difficult for the sector to manage client knowledge and service quality. Numerous challenges persist in the industry, highlighting the ongoing need for continuous learning and research in Malaysia to enhance the management of courier and postage services, making them more competitive. In ensuring the sustainability of all courier service providers, maintaining a strong focus on customer satisfaction remains paramount. This requires a collective and concerted effort to prioritize dimensions such as service quality and customer knowledge within the industry.

3 Customer Satisfaction

At its core, customer satisfaction encapsulates customer testimonials, whether positive or negative, reflecting the attributes of a product or the offerings of a service [7]. Additionally, it assesses the degree of client satisfaction with the overall functionality, services, and offerings of a business [8]. A nuanced understanding of customer satisfaction involves delineating it into two fundamental aspects: product organization and service organization. This distinction is pivotal, providing organizations with a framework to formulate quality strategies, concentrating on enhancing both products and services to elevate overall customer satisfaction.

Customer satisfaction dictates a pivotal role in the operations of businesses and organizations spanning diverse sectors, exerting a profound impact on a spectrum of outcomes, both positive and negative. Favorable customer satisfaction outcomes manifest in the cultivation of client devotion and generation of favorable rumors [9], the stimulation of repeat purchases [11], the influence on customer acquisition and retention dynamics [12], and the establishment of customer trust [10]. Conversely, adverse customer satisfaction results can lead to undesirable consequences.

Within the courier services industry, assessing customer satisfaction is vital for courier company management to understand the factors impacting customer contentment. This knowledge can present advantages over competitors and uncover opportunities. It is essential for these companies to grasp customer needs and expectations, emphasizing key dimensions of service quality like responsiveness, reliability, empathy, tangibility, and assurance [3]. Crucial processes in courier services, affecting customer satisfaction, involve meeting delivery expectations on time and minimizing delays and lost items.

4 Service Quality

Broadly speaking, service quality refers to the general excellence and performance of a service from the perspective of customers. It encapsulates a service provider's ability to efficiently satisfy customers, thereby contributing to an improved delivery process [12]. Although service quality is a multifaceted notion, its core elements consists that of assurance, tangibility, empathy, reliability, and responsiveness, functioning for the basis for evaluating the rendered service quality [13]. In the context of mass service, such as in logistics overall and specifically in courier services, Nevertheless, it will be elaborated further that, specifically for courier services, the dimensions of service quality designed to be relevant are tangibility and competence, while most studies commonly employ the five categories of service quality, there is a necessity in fine-tuning the service quality platforms to align with diverse contexts [14].

5 Knowledge for Customer

The effectiveness of courier services for users is closely tied to the knowledge base, suggesting that customers find greater satisfaction when service industry employees, particularly in the mass service context like the courier services sector, possess extensive knowledge. In this context, customer knowledge includes knowing how courier services operate, knowing enough about how much they charge, being comfortable with their tools, and being aware of their routes. All these factors contribute to customer comfort and satisfaction. The dimensions of information needs and customer education are critical to meeting the knowledge requirements necessary for understanding the delivery process in the courier services business within the framework of the knowledge for customer construct. Customers naturally seek sufficient information before making purchases, as it significantly influences their choices. The function of addressing information needs aids customers in decreasing ambiguity throughout the phase of decision-making. Customers typically continue seeking relevant and reliable information until their knowledge gap is narrowed, and they acquire new knowledge. Furthermore, efforts invested in educating customers are regarded as a proposal of worth within the service process. This serves as a favorable indication of the organization's obligation to maintaining good rapport with customers.

6 Methodology

This study investigated customer happiness, service quality, and customer knowledge in the courier services business through practical fieldwork utilizing a quantitative technique and a self-administered questionnaire. The survey instrument was the questionnaire and the research adopted a descriptive research design. Initial research efforts were concentrated in the Johor Bahru district, targeting existing customers utilizing diverse courier service companies. Data analysis involved both descriptive and inferential methods, with a convenient sampling technique appropriately applied. In summarizing customer satisfaction, service quality, and knowledge for customers, Fig. 1 illustrates their relationship. In this conceptual framework, customer satisfaction is the dependent variable, service quality is the independent variable, and knowledge for the customer is the mediating variable, outlining the distinct roles of each construct.

7 Results

Concerning the constructs that interconnect the variables, the assessment of the structural model or inner model involved assessing its predictive capabilities and the associations between the constructs. In this study, the structural model is depicted

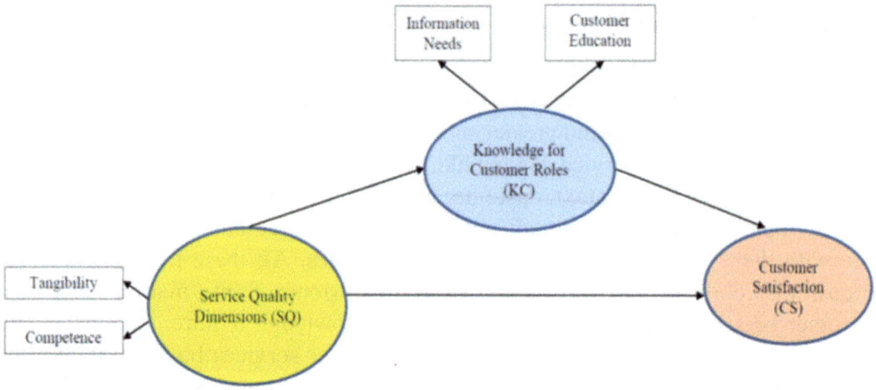

Fig. 1 Conceptual model

in Table 2, illustrating the independent latent variable, dependent latent variable, and mediating variable. This calculation encompasses all the constructs connected to each other to formulate a model of a research. Two specifications determine the hypothesized relationships between constructs in the study: the coefficient of determination (R^2) of endogenous constructs and the path coefficients of hypothesized relationships. Whether the external constructs influence the endogenous constructs will be shown by the assessment's outcomes.

The findings from the present study reveal that R^2 value for customer satisfaction is 0.718, customer education is 0.556, and information needs is 0.52. Notably, the highest R^2 value is attributed to customer satisfaction, reaching 0.718. This implies that 71.8 percent of the variability observed in customer satisfaction can be elucidated by the independent variables under consideration. Conversely, the path coefficients are employed to assess the magnitude and statistical importance of the proposed associations among latent constructs within the structural model. The estimations yield standardized values extending from -1 to $+1$, wherein a coefficient closer to $+1$ signifies a robust positive relationship, while a coefficient approaching -1 designates a substantial negative relationship. The direct relationships depicting these coefficients are detailed in Table 3.

By using the indirect effect analysis procedure, the results regarding the mediation influence of customer knowledge in the association of service quality and customer satisfaction will be clarified. A detailed breakdown of knowledge for customer will

Table 2 Coefficient of determination (R^2) of endogenous constructs

Constructs	R^2 adjusted
Customer satisfaction (CS)	0.718 0.714
Customer education (CE)	0.556 0.554
Information needs (IN)	0.520 0.517

Table 3 Hypothesis testing (direct effect)

Hypothesis	Path	Path coefficient	Std error	*t*-value	*P* value	Decision
H1a	Tan → CS	0.225	0.062	3.582	0.000	Yes
H1b	Comp → CS	0.144	0.055	2.611	0.000	Yes
H2a	Tan → IN	0.362	0.077	4.620	0.000	Yes
H2b	Tan → CE	0.264	0.071	3.702	0.000	Yes
H2c	Comp → IN	0.368	0.094	3.894	0.000	Yes
H2d	Comp → CE	0.581	0.064	8.972	0.000	Yes
H3a	IN → CS	0.241	0.075	3.203	0.000	Yes
H3b	CE → CS	0.146	0.054	2.693	0.000	Yes

serve as mediating variables [15, 16]. The mediation link involving knowledge for customer, service quality and customer satisfaction is explicated in Table 4.

The findings suggest that customer satisfaction in the courier services business is directly impacted by the quality of the services provided. The results imply that within the courier services sector, service quality has a direct effect to customer satisfaction, and the knowledge for the customer construct partially mediates this relationship. Positive correlations exist, especially in the tangibility and competence dimensions of service quality, demonstrating their influence on customer satisfaction. Tangibility and competence also positively correlate with the knowledge for the customer construct, emphasizing the role of service quality in imparting sufficient customer knowledge. Additionally, elements like information needs and customer education within the knowledge for the customer construct positively influence customer satisfaction, partially mediating the service quality-customer satisfaction connection.

Table 4 Mediation type

Hypothesis	Path	*t*-value	*P* value	Direct effect with mediator	Direct effect without mediator	Mediation type
H4a	Tan → IN → CS	3.916	0.000	0.240**	0.367**	Partially
H4b	Comp → IN → CS	2.965	0.003	0.228**	0.359**	Partially
H4c	Tan → CE → CS	3.254	0.001	0.287**	0.367**	Partially
H4d	Comp → CE → CS	4.810	0.000	0.182**	0.359**	Partially

8 Conclusion

Customer satisfaction relies on efficient service delivery, especially in meeting continuous information needs. Customer service personnel grapple with imparting knowledge to increase the product and service's value. Prioritizing knowledge transfer in the courier services industry signifies a business commitment to elucidating logistics processes. For courier service providers, delivering top-notch services and effective knowledge contributes substantial value, deepening customer understanding and fostering heightened awareness of overall courier company operations.

References

1. T.Y. Kin, A.H. Suzanawati, Z. Siti Nur Atikah, Data envelopment analysis for Malaysia courier services performance and customer satisfaction. J. Comput. Res. Innov. **5**(3), 25–33 (2020)
2. P.S. Liew, W.N. Liew, C.S. Lim, J.Y. Lim, S.I. Qonitah, The impact of Covid-19 pandemic on consumer behavior: J&T Express courier services in Malaysia and Indonesia. Asia Pac. J. Manag. Educ. **5**(1), 65–78 (2022)
3. R.Q. Tang, Y.J. Tan, Z.X. Tan, Y.T. Tan, G. Almawad, A. Alosaimi, A study of courier service quality and customer satisfaction. Int. J. Appl. Bus. Int. Manag. **7**(1), 137–150 (2022)
4. F. Siali, A.W.S. Wen, M.U.A. Hajazi, Booming of online shopping in Malaysia: Do customers satisfy with parcel delivery service? Int. J. Acad. Res. Bus. Soc. Sci. **8**(12), 415–436 (2018)
5. N. Izzah, F. Dilaila, L. Yao, The growth of reliance towards courier services through e-business verified during COVID-19, Malaysia. J. Phys. Conf. Ser. **1874** (2021)
6. N. Izzah, F. Dilaila, L. Yao, Relationship—courier partner logistics and e-commerce enterprises in Malaysia: A review, Indian. J. Sci. Technol. **9**(9), 1–10 (2016)
7. S. Khairawati, Effect of customer loyalty program on customer satisfaction and its impact on customer loyalty. Int. J. Res. Bus. Soc. Sci. **9**(1), 15–23 (2020)
8. J.T. Libo-on, Service quality influence on customer satisfaction in courier services: A comparative study. Am. Int. J. Bus. Manag. **4**(3), 51–63 (2021)
9. D. Darmawan, The effect of customer satisfaction on trust and customer loyalty. Manag. Account. Res. J. **3**(2), 1–73 (2019)
10. D. Choi, C.Y. Chung, J. Young, Sustainable online shopping logistics for customer satisfaction and repeat purchasing behaviour Evidence from China. Sustainability **11**(20), 1–20 (2019)
11. M. Sağlam, S. El Montaser, The effect of customer relationship marketing in customer retention and customer acquisition. Int. J. Commer. Financ. **7**(1), 191–201 (2021)
12. N. Ramya, A. Kowsalya, K. Dharanipriya, Service quality and its dimensions. EPRA J. Res. Dev. **4**(2), 38–41 (2019)
13. V.A. Zeithaml, L.L. Berry, A. Parasuraman, The behavioural consequences of service quality. J. Mark. **60**, 31–46 (1996)
14. M. Kumar, F.T. Kee, V. Charles, Comparative evaluation of critical factors in delivering service quality of banks: An application of dominance analysis in modified SERVQUAL model. Int. J. Qual. Reliab. Manage. **27**(3), 351–377 (2010)
15. C. Nitzl, J.L. Roldan, G. Cepeda, Mediation analysis in partial least square path modelling: Helping researchers discuss more sophisticated models. Ind. Manag. Data Syst. **116**(9), 1849–1864 (2016)
16. X. Zhao, J.G. Lynch Jr., Q. Chen, Reconsidering baron and Kenny: Myths and truths about mediation analysis. J. Consum. Res. **37**(2), 197–206 (2010)

Impact of Covid-19 Toward Consumer's Viewpoint in Food Delivery Services

Mazlan Md Zahid, Aminah Syakirah Ahmad Topik, Mohd Nasir Alias, Hairul Rizad Md Sapry, and Mustakim Melan

Abstract The provision of food delivery services through online platforms has emerged as a fundamental requirement of contemporary society. This is due to the fact that the process of placing an order online is more time-efficient compared to the alternative of physically visiting a store to place an order. The present circumstances are characterized by the prevalence of the Coronavirus-19 pandemic, which necessitates adherence to governmental protocols and limits individuals' freedom of movement. This research investigates the influence of the COVID-19 pandemic on consumers' perceptions of the competencies of the foodpanda application's food delivery service. The study aims to provide recommendations based on the feedback received from consumers. A quantitative approach was employed to gather data, whereby a series of questionnaires were disseminated to a sample of 100 individuals utilizing the application residing in Pasir Gudang, Johor. The research is centered on two distinct variables that are not interdependent, namely customer service and product of promotion. The study revealed that customer service and promotional product offerings are significant factors in the research. In summary, the researchers hope that the proposed recommendations will prove beneficial to users of the application, resulting in the retention of current consumers and the attraction of potential users in the future.

M. M. Zahid (✉) · A. S. A. Topik · M. N. Alias · H. R. M. Sapry
Industrial Logistics, Malaysian Institute of Industrial Technology, Universiti Kuala Lumpur, Masai, Malaysia
e-mail: mazlanmz@unikl.edu.my

A. S. A. Topik
e-mail: aminah.topik@s.unikl.edu.my

M. N. Alias
e-mail: mohdnasir@unikl.edu.my

H. R. M. Sapry
e-mail: hairulrizad@unikl.edu.my

M. Melan
Universiti Utara Malaysia, Sintok, Malaysia
e-mail: mustakim@uum.edu.my

A. Ismail et al. (eds.), *Industrial Logistics I*,
SpringerBriefs in Applied Sciences and Technology,
https://doi.org/10.1007/978-3-032-26981-2_13

Keywords Consumer's viewpoint · Customer service · Product of promotion

1 Introduction

Since Covid-19 take place in Malaysia, the food delivery service is well served with numbers of orders they received and not to mention, the technology in food delivery services is also improved. Many restaurants, eateries, and cafes took this opportunity to maximize their sales. This research focuses on the food delivery service that has been chosen, which is foodpanda which has not reached to the expected level of satisfaction. Along with the introduction of Internet technology, certain prominent fast-food businesses, especially pizza restaurants, have been the pioneers to adopt online food ordering with their websites. The adoption of online meal ordering by restaurants can be attributed to the fact that it has fulfilled or exceeded expectations in a number of areas for restaurant operations [1]. Oltra -Mestre et al. [2] further explained that because of the many advantages it offers, customers and restaurants are increasingly turning to the option of placing orders for meals online. According to [2], several novel ideas were included into the system of placing food order online through various websites. According to the findings of A Study on Customer's Attitude and Perception Towards Digital Food App Services [3], services are intangible products since there is no way to transfer possession or ownership of them, they cannot be sold, and they do not come into existence until after they have been purchased or consumed by a customer.

The contentment of the customers depended on a variety of aspects, including the accessibility of the meal, the variety of payment options, the customer ratings, and the amount of human connection [4]. The primary objective of food delivery services should be to achieve maximum client satisfaction rather than simply to maximize profits. Service providers need to place a primary emphasis on the quality of their services in order to achieve maximum levels of customer happiness [5]. In the fast-paced world of today, where customers' busy schedules mean time is in short supply, time-saving orientation, also known as TSO, has become an essential component in easing daily duties while making the most of available time. Many people who worked in offices were unable to spare the time and effort required to go out to restaurants, which typically involved driving there and waiting in line to place their orders. Therefore, utilizing OFDS is the shortest way to acquire food, and the time saved can be put toward the completion of other responsibilities. Because of the opportunity costs, customers with higher incomes place a higher value on their time. As a result, people find tempting the option of buying online because it enables them to make more efficient use of their time [6].

According to [7], convenience for the consumer is paramount in the food delivery services system's essential qualities. It is crucial to market their business and the services they provide to clients by advertising their offering, particularly on the customer's request, to make it easier for customers to place orders with them. The importance of focusing on providing high-quality service that prioritizes the needs

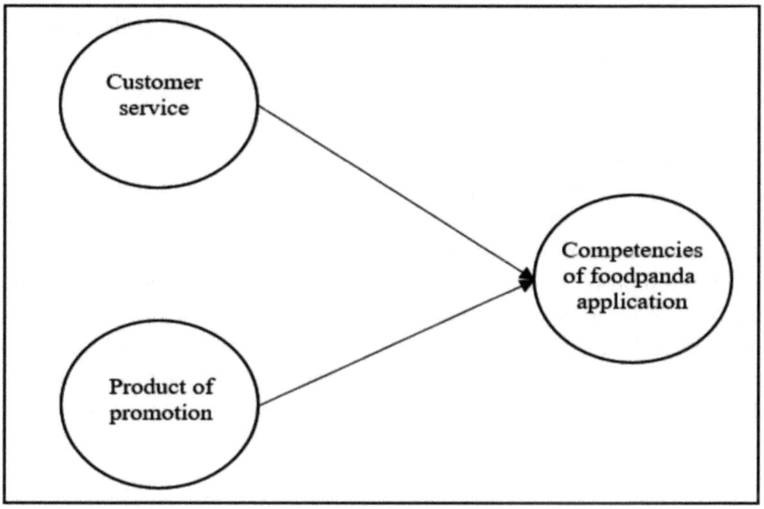

Fig. 1 Conceptual framework

of customers cannot be over-stated. Jimenez et al. [8] found that only a few studies explain customer experience issues with online food delivery services and variables such as convenience, hedonism promotion, association to save money, time-saving, customer experience purchase intention, consumer behavior and behavioral intention to influence customers when they search for food delivery service online service application.

Figure 1 shows the conceptual framework of the research study which relates to the consumer's viewpoint about competencies of foodpanda application. By looking, it gives an overview of the framework implements structured phases and provides a better understanding of the variables used. The factors that affect the user's viewpoint of the food delivery application are customer service and products of promotion. These factors are essential to identify because they will determine the outcome of the study is going well and on track. In addition to this, the factors will make stronger its standing by stating a few sub factors that will be supporting the main one.

2 Methodology

In this study, a quantitative method has been applied. The researcher used a survey to obtain data from respondents by using Google forms and send it online to all users of the application at Masai, Johor. The collection of quantitative information enables researchers to perform simple to very sophisticated statistical analyses such as aggregating data, showing relationships between data, and comparing aggregated data. Yeo et al. [9] stated that qualitative research involves the collection and analysis of narratives and/or open-ended observations through methodologies such as interviews, focus groups or ethnographies. The researcher believes that the findings that will be conducted from a small group of individuals to a large one can be generalized using quantitative research. The method is structured and statistical, helping researchers complete their analyses and make informed decisions from previous conclusions. Research methodology is a method to define how the researcher plans to conduct the investigation of the research. It is one of the systematically plans to resolve a research problem. A methodology also describes the researcher's approach to the study in order to ensure reliable and the valid results that address the objective of the research. The researcher uses a quantitative research technique which is concerned with quantifying and analyzing variables in order to get data. A good methodology will help the researcher to find the answers needed in the research question and achieve the research objective. This study will employ the qualitative research method.

The researcher has chosen a population of users in Masai, Johor, that uses the food delivery service, foodpanda application. The number of respondents was decided by sampling to get the specific sample size. It is essential to accomplish this in order to collect responses from legitimate and accurate respondents to the questionnaire. This part was concerned with the customer service in the foodpanda application that are important to achieve user's level of satisfaction. Five scales were used in the questionnaire to answer questions.

Primary data are the designed instruments that are used in this study, enabling the researcher to obtaining data directly from the sources. The questionnaires are created by three sections: demographic, customer service, and product of promotion as per Table 1.

Closed-ended questions are applied. Questions offer respondents a fixed set of choices to select from. The respondent will be asked to choose yes or no or select from multiple-choice answers. Other than that, the researcher also chose a Likert scale 1–5 because it is easy to collect the data and not give the respondents thinking

Table 1 Question design

Section demographic index
This part focused on demographic data such as respondent age, gender, ethnicity, status, and income. Close-ended questions were used in the questionnaire to respond the questions.
Section customer service index
This part was concerned with the customer service in the foodpanda application, which is important to achieve consumers' level of satisfaction. Close-ended and five scales were used in the questionnaire to answer questions.
Section product of promotion
This part was concerned with the product of promotions shown in the foodpanda application that are able to achieve consumer's level of satisfaction. Close-ended and five scales were used in the questionnaire to answer questions.

Table 2 Likert scale satisfaction measurement

1	2	3	4	5
Strongly Disagree	Disagree	Neutral	Agree	Strongly agree

Source Bhandari (2021) Questionnaire design

the questionnaire that was distributed to the respondents is hard to answer as per Table 2.

3 Results and Discussion

In Table 3, for customer service, 59 out of 100 respondents have never contacted foodpanda's customer service center during Covid-19 period compared to other 41 respondents who have contacted. Furthermore, for the question of do they find a solution regarding their problem after contacted with the foodpanda's customer service center gained 58 positive answers, which is they found a solution after the session. For the statement of their satisfaction with the service, it shown the highest answer is neutral and second highest is agreeing, which mean that they are satisfy with the service given. From this, it indicates that their customer service center is quite handy and reliable.

In section product of promotion, 62% of the respondents use promotion code 1–5 times a month, followed by second highest with 29% with none of them are using it

Table 3 Section customer service

<table>
<tr><th rowspan="2">Question</th><th colspan="2">Respondents (%)</th></tr>
<tr><th>Yes</th><th>No</th></tr>
<tr><td>Have you ever contacted with the foodpanda's customer service center throughout Covid-19 period?</td><td>41</td><td>59</td></tr>
<tr><td>Do you find a solution regarding your problem after contacted with the foodpanda's customer service center?</td><td>58</td><td>42</td></tr>
</table>

<table>
<tr><th rowspan="2">Question</th><th colspan="3">Respondents (%)</th></tr>
<tr><th>None</th><th>1–5 times a month</th><th>6–10 times a month</th></tr>
<tr><td>How frequent you contacted with the foodpanda's customer service center throughout Covid-19 period?</td><td>58</td><td>41</td><td>1</td></tr>
</table>

<table>
<tr><th rowspan="2">Variables</th><th colspan="5">Respondents (%)</th></tr>
<tr><th>Strongly disagree</th><th>Disagree</th><th>Neutral</th><th>Agree</th><th>Strongly agree</th></tr>
<tr><td>You are satisfied with the service that foodpanda's customer service center provided</td><td>3</td><td>2</td><td>53</td><td>32</td><td>10</td></tr>
</table>

Table 4 Section product of promotion

Question	Respondents (%)			
	None	1–5 times a month	6–10 times a month	11–15 times a month
Howfrequentyou use promotion code (like picture below) throughout Covid-19 period?	29	62	7	22

Variables	Respondents (%)				
	Strongly disagree	Disagree	Neutral	Agree	Strongly agree
You are satisfied with the service of promotion code that foodpanda has provided	2	4	39	42	13

Question	Respondents (%)	
	Yes	No
Do you face any problem while using the promotion code that foodpanda has provided?	82	18

during Covid-19. Next, with 22 of 100 respondents using it 11–15 times a month and lastly, only 7% apply the promotion code 6–10 times a month. Based on the "You are satisfied with the service of promotion code that foodpanda has provided", almost half of the respondent agree that they are satisfied with the promotion code given. It shows that those promotion codes are quite reasonable and useful to the consumers (Table 4).

4 Conclusion

Consequently, individuals are compelled to either prepare their own meals or utilize online food delivery services such as foodpanda and Grab. It is possible for orders to reach a maximum of 100 at a given time, and the individuals responsible for meal preparation are human, thus susceptible to human error, such as omitting an item or under-seasoning the food. Upon detecting an issue, customers typically contact the customer service center to obtain assistance. By engaging a competent representative to address the issue, all challenges can be resolved seamlessly without any undesirable occurrences. Nevertheless, removing the invalid promotion code can avoid dissatisfied consumers. Elicited a response from the majority of participants indicating that they experienced issues with the provided promotion code. These issues were primarily related to the code reaching its maximum limit and being deemed invalid. To prevent consumer dissatisfaction, foodpanda could expeditiously remove any promotional codes that have either expired or reached their limit of redemption.

References

1. Y. Chen, S. Fay, Q. Wang, The role of marketing in social media: how online consumer reviews evolve. J. Interact. Market. **25**(2), 85–94 (2011)
2. M.J. Oltra -Mestre, V. Hargaden, P. Coughlan, B. Segura-García del Río, Innovation in the agri-food sector: exploiting opportunities for industry 4.0. Creat. Innov. Manag. **1**, 198–210 (2020)
3. K. Mohamed Anwar, D. Vijayalakshmi, Result of psychological well-being on online food delivery service executives job performance. J. Comput. Theor. **12**, 1120–1130 (2019)
4. A. Nicolaides, *Service quality, empowerment and ethics in the South African hospitality and tourism industry and the road ahead using ISO9000/1* (2008). https://uzspace.unizulu.ac.za/server/api/core/bitstreams/19af17e7-2e31- 41b2-bbac-0010754fca21/content
5. S.E. Kimes, *The Current State of Online Food Ordering in the U.S. Restaurant Industry* (2011). https://hdl.handle.net/1813/71075
6. J. Suganya Devi, J. Sakthi Jayalakshmi, G.Monisha, Outsourcing to online food delivery services: perspective of F&B business owners. EPRA. Int. J. Res. Development. **22**(2), 40–43 (2019)
7. K. Martins-Turnera, A. Grahleb, K. Nagela, D. Gohlich, Electrification of urban freight transport—a case study of the food retailing industry. Procedia. Computer. Sci. **170**, 757–763 (2020)
8. D. Jimenez, S. Valdes, M. Salinas, Popularity comparison between e-commerce and traditional retail business. Int. J. Technol. Bus. **1**(1), 10–16 (2019)
9. V.C.S. Yeo, S.K. Goh, S. Rezaei, Consumer experiences, attitude and behaveioral intention toward online food delivery (OFD) services. J. Retail. Consum. Serv. **35**(C), 50–162 (2017)

Using the E-servqual Model for Measuring E-Hailing and Taxi's Service Quality in Pasir Gudang, Johor.

Mazlan Md Zahid, Noor Shazwa Hajeerah Adnan, Mohd Nasir Alias, and Mustakim Melan

Abstract Ride-hailing services are defined as booking transportations service of private vehicles through mobile applications in association with transportation network providers to publics. E-hailing and taxi services were included in the ride-hailing services as it offers public transportation service for private vehicles. Recently, the growth of E-hailing services was increasing compared to taxi services. Thus, the decreasing demand for taxi services needs to be investigated. Quality assurance is important to be measured in order to make a parallel comparison for the services offered. This paper analyzed the use of the E-servqual model containing of reliability, assurance, tangibility, empathy, and responsiveness to construct E-Service quality for E-hailing and taxi's service industry. A quantitative method is implemented to measure the difference of satisfaction among passengers. Hence, a total of 191 respondents were involved in this study. Researcher discovered an assurance was one of the important factors of service quality that affects customer's satisfaction for E-hailing and taxi's services industry.

Keywords Ride-hailing · E-hailing · Taxi service · E-Servqual model · Service quality

---

M. M. Zahid (✉) · N. S. H. Adnan · M. N. Alias
Industrial Logistics, Malaysian Institute of Industrial Technology, Universiti Kuala Lumpur, Johor, Malaysia
e-mail: mazlanmz@unikl.edu.my

N. S. H. Adnan
e-mail: shazwa.adnan@s.unikl.edu.my

M. N. Alias
e-mail: mohdnasir@unikl.edu.my

M. Melan
Universiti Utara Malaysia, Sintok, Malaysia
e-mail: mustakim@uum.edu.my

A. Ismail et al. (eds.), *Industrial Logistics I*,
SpringerBriefs in Applied Sciences and Technology,
https://doi.org/10.1007/978-3-032-26981-2_14

1 Introduction

E-hailing is a public transportation service that offers online booking through electronic applications and it was known since 2014 and it gives significant impact to the growth of transportation services in Malaysia. The rapid development of Internet and technology had encouraged the development of online booking of public transportation for private vehicles. Before the existence of E-hailing, taxi services were the most well-known public transportation for private vehicles. Due to the issue arise with the unbalanced demand for E-hailing and taxi industry in Malaysia, this research aims to investigate the service quality and customer satisfaction of E-hailing and taxi industry by using the E-servqual model. The E-servqual model proposed by [1] which uses a multiple item-scale in measuring the consumer perceptions for service quality, including the five dimensions such as the reliability, assurance, tangible, empathy and responsiveness. This research is significant to undertake a parallel comparison of E-service quality perception between E-hailing and taxis service industry in Malaysia.

1.1 Reliability

According to [2], reliability has a significant and major impact on customer satisfaction in Ghanaian minicab taxi services. Alike [3] and [4], both showed that a public transportation company's reliability is strongly correlated with customer's satisfaction for each services. A study conducted in New Delhi by [5] found that attractive fares discounts, no parking to find or pay, easy service request, and real-time data exchange be an important parameter to satisfy the ride-hailing passengers.

1.2 Responsiveness

Based on the study made by [6], responsiveness can be identified as the primary determinant of service quality in Kumasi transportation systems. Passenger satisfaction is favorably impacted by the responsiveness of the service offered. As a results, the service provider should be dedicated to provide a prompt service response, being accessible to passengers at all times and informed them the status of the services. The ride-hailing drivers must be sincere and responsive in delivering their services. Examples include offering a timely schedule, keeping passenger informed, and respond each of their questions or complaints. As a result, the responsiveness is frequently recognized as a key enabler in the service sector.

1.3 Tangibility

Based on the study made by [7], the cleanliness of the ride-hailing vehicles is one of the important factors in tangibility which achieve customer satisfaction. Vehicle cleanliness is the most essential element in determining the passenger's satisfaction of using the ride-hailing services. A study in Hong Kong found that the professionalism and attractiveness attributes of services such as professionalism of drivers were the significant factors that are influencing customer's satisfaction of the services offered. According to [8], the service convenience and vehicle cleanliness, comfort and quality have been proved to be a key factor in influencing of how users rate the service quality that they perceived.

1.4 Assurance

Accordingly, in achieving customer satisfaction involves the driver's skill and knowledge to enhance customer confidence in the ride-hailing services. In addition, the knowledge and expertise of the driver give passenger a sense of security, which enable them to have a peace of mind, and assurance that the driver will transit them to the requested destination on time. Furthermore, [9] studied the influence of service quality on Malaysian's user satisfaction with ride-hailing services. According to the empirical data, customer satisfaction is significantly influenced by assurance. The level of satisfaction among passengers arise when the driver is able to deliver service that can be relied upon, such as reliable transportation that can get them to the requested destination safely.

1.5 Empathy

Aziz et al. [10] mentioned that empathy is defined as the employees' behavior toward the customer based on their emotional experiences and observations were also discovered to be significantly and positively connected to customer satisfaction. According to the [11], there is a significant impact of service quality when the driver understands the passengers needs and offers services which give more convenient for them. As a result, it generates a bond with the passengers and offers services that come close to satisfy their needs, thus, enhance their satisfaction.

Research Objectives:

1. To investigate the factors of service quality such as reliability, responsiveness, tangibility, assurance, and empathy for E-hailing and taxi's services industry.
2. To study the comparison of service quality between E-hailing and taxi's services industry.

2 Methodology

Research methodology is a method to define how the researcher plans to conduct the investigation of the research. It is one of the systematically plans to resolve a research problem. A methodology also describes the researcher's approach to the study in order to ensure reliable and the valid results that address the objective of the research. The researcher uses a quantitative research technique which is concerned with quantifying and analyzing variables in order to get data. A good methodology will help the researcher to find the answers needed in the research question and achieve the research objective. This study will employ the qualitative research method.

The researcher has chosen a population in Pasir Gudang, Johor, specifically in Bandar Seri Alam, Masai. Students in Unikl Mitec and Uitm Pasir Gudang had been selected as the population of this study who frequently use the E-hailing and taxi services in their daily basis. The purpose of this study is to identify the factors of E-servqual model (E-service quality) that affects the customer satisfaction of the users for E-hailing and taxi services. Thus, it is important to choose the correct population in order to get the valid and correct answers from the respondents. In this study, the number of the targeted population in Universiti Kuala Lumpur (MITEC) and Universiti Teknologi MARA (UiTM) Pasir Gudang, Johor is 3411 in total which involve 1811 students for UniKL Mitec and 1600 students for UiTM Pasir Gudang, Johor. Thus, the recommended sample size calculated by using Raosoft sample size calculator was 160 (margin of error=7%, confident level=93). Therefore, a total of 160 Google Form are distributed to the targeted respondents.

3 Results and Discussion

The interest was to find the answers for which factors of service quality, such as reliability, responsiveness, tangibility, assurance, and empathy impact the ride-hailing services which is E-hailing and taxi's services industry.

The highest percentage among the five factors of E-servqual model for E-hailing services (see Table 1) is 49.20% which involves the factors of assurances. A total of 94 from 191 respondents choose very satisfied with the service quality offered by E-hailing which is ensuring all passenger reach the destination on time and safely compared to the other two assurance service quality which is E-hailing drivers are trustworthy and arrive on time to pick-up passengers.

The highest percentage among the five factors of E-servqual model for taxi services (see Table 2) is 46.1% which involves the factors of assurances. The total of 88 from 191 respondents answer neutral with the service quality offered by taxi services that mentioned taxi drivers are trustworthy. The neutral Likert scale here may not convey the real answers for the research objectives. In order to answer the objectives of the research, the researcher choose the second highest percentage from the overall results for taxi services.

Table 1: Analysis 1

Variables	Respondents (%)				
	Not at all satisfied	Slightly Satisfied	Neutral	Very satisfied	Extremely satisfied
A. Reliability					
E-hailing (Grab-rides, Maxim, In-Driver, and Air Asia, etc.) providing services as promised	0	6.30	23.60	**46.60**	23.60
E-hailing service fare is reliable	3.10	18.30	30.90	**26.20**	21.50
Security aspect for E-hailing is satisfactory	2.60	8.40	24.10	**45.00**	19.90
B. Responsiveness					
E-hailing provides timely and efficient service	0	7.90	23.60	**44.50**	24.10
E-hailing drivers keep the customer informed of any delay or inability to deliver the services	0.50	9.90	22.50	39.80	27.20
Communication with E-hailing drivers is clear and helpful	0.50	4.20	24.10	42.40	28.80
C. Tangibility					
E-hailing driver's attire is neat	0	7.90	35.60	38.70	17.80
The interior of E-hailing vehicle is clean	1.60	6.80	21.50	40.80	29.30
E-hailing application provides detail information	0.50	3.10	19.40	**44.50**	32.50
D. Assurance					
Drivers are trustworthy	1	4.70	33.50	**42.40**	18.30
Driving ensures passenger reach destinations on time and safely	0	3.70	22.50	**49.20**	24.60
Drivers arrive on time to pick-up passengers	1	9.40	28.30	41.90	19.40
E. Empathy					
Drivers understand customer's needs and wants	2.10	5.20	**36.10**	36.10	21.50
Drivers seldom reject booked ride	2.10	15.70	30.90	31.90	19.40
E-hailing provides 24/7 services	20.90	35.10	29.30	9.40	5.20

It involves responsiveness factors with 45% which includes 86 respondents answered slightly satisfied. The majority of the respondents choose slightly satisfied with the taxi drivers that keep the customer informed of any delay or inability to deliver the services compared to the other two responsiveness service quality that get lower percentage which is taxi's services provide timely and efficient service and communication with taxi drivers are clear and helpful.

Table 2: Analysis 2

Variables	Respondents (%)				
	Not at all satisfied	Slightly satisfied	Neutral	Very satisfied	Extremely satisfied
A. Reliability					
1. Taxi provides services as promised	1.60	28.80	41.90	25.10	2.60
2. Taxi service is reliable	12.60	42.90	33.00	8.40	3.10
3. Security aspects for taxi services are satisfactory	4.70	36.10	40.80	14.10	4.20
B. Responsiveness					
1. Taxi services provide timely and efficient service	3.10	41.90	35.60	16.20	3.10
2. Taxi drivers keep the customer informed of any delay or inability to deliver the services	5.80	45.90	34.40	13.60	1.60
3. Communication with taxi drivers are clear and helpful	2.10	34.60	38.70	22.50	2.10
C. Tangibility					
1. Taxi driver's attire is neat	2.10	31.40	34.60	25.70	6.30
2. The interior of taxi vehicle is clean	1.60	31.90	33.50	28.30	4.70
3. Taxi application provides detail information	26.70	25.10	31.90	12.00	4.20
D. Assurance					
1. Taxi drivers are trustworthy	2.60	27.20	46.10	21.50	2.60
2. Taxi drivers ensure passenger reach destination on time and safely	2.10	25.90	41.40	27.70	3.10
3. Taxi drivers arrive on time to pick-up passengers	2.10	34.60	37.70	22.00	3.70
E. Empathy					

(continued)

Table 2: (continued)

Variables	Respondents (%)				
	Not at all satisfied	Slightly satisfied	Neutral	Very satisfied	Extremely satisfied
1. Taxi drivers understand customer's needs and wants	1.60	36.60	35.10	24.60	3.10
2. Taxi drivers seldom reject booked rides	2.10	35.10	40.30	19.40	3.10
3. Taxi services provide 24/7 services	28.80	28.60	27.20	11.50	3.70

4 Conclusion

The researcher has accomplished both research objectives. Firstly, is to investigate the factors of service quality such as reliability, responsiveness, tangibility, assurance and empathy for E-hailing and taxi's services industry. The findings have indicated that assurances are the factor of service quality for E-hailing and taxi services. For E-hailing services, it indicates that customers are very satisfied with the factors of service quality of assurance. For taxi services, the highest results involved with neutral Likert scale in which it does not represent the real objectives of the research. Thus, the second highest result indicates that customers are slightly satisfied with the factors of service quality of responsiveness. The second research objective is to compare the service quality between E-hailing and taxi's services industry. The finding has indicated that the majority of the customer are very satisfied with the service quality of E-hailing drivers that ensure passenger reach destinations on time and safely while the majority of customer felt neutral with the service quality of taxi drivers are trustworthy. The second highest result of service quality for taxi is involved with customers choose slightly satisfied with the taxi's drivers that keep the customer informed of any delay or inability to deliver the services. Thus, this research measured the service quality of both ride- hailing services by using the E-servqual model which absolutely gives impact toward the customer satisfactions and the growth of demand for both services.

References

1. A. Parasuraman, V.A. Zeithaml, L. Berry, SERVQUAL: a multiple-item scale for measuring consumer perceptions of service quality. **64**(1), 12–40 (1988). https://psycnet.apa.org/record/1989-10632-001
2. N. Horsu, P. Yeboah, Influence of service quality on customer satisfaction: a study of minicab taxi services in cape coast. Ghana. **3**, 1451 (2015). http://ijecm.co.uk/ISSN23480386

3. M. Khuong, Q. Dai, The factors affecting customer satis faction and customer loyalty—a study of local taxi companies in Ho Chi Minh City, Vietnam. **7**, 228–233 (2016). https://www.ijimt.org/678-MB00013
4. Olawole et al., An empirical study of commuters' satisfactions with taxi service quality in Abeokuta, Nigeria. Transport. Res. Interdiscipl. Perspect. **11**, 1–8. https://doi.org/10.1016/j.trip.2021.100434
5. Urban Transport in the Sharing Economy Era, Buenos Aires: CIPPEC (2018)
6. Sam et al., SERVQUAL analysis of public bus transport services in Kumasi metropolis, Ghana: Core user perspectives. Case Stud. Transp. Policy. **6**, 25–31 (2017). https://doi.org/10.1016/j.cstp.2017.12.004
7. Tang et al., A study of courier service quality and customer satis faction. Int. J. Appl. Bus. Int. Manage Ment. **7** (2022). https://ejournal.aibpmjournals.com/index.php/IJABIM/article/view/1447
8. Wong, Szeto., Modeling the acceptance of taxi owners and drivers to operate premium electric taxis: Policy insights into improving taxi service quality and reducing air pollution. Transport. Res. Part A: Policy Pract. **118**, 581–593 (2018). https://www.sciencedirect.com/science/article/abs/pii/S0965856418303707
9. Evaluation of Service Quality Dimensions towards Customer's Satisfaction of Ride-Hailing Services inKualaLumpur, Malaysia. http://www.researchgate.net/publication (2019). Accessed 2 February 2023
10. Aziz et al., Effect of employee empathy on customer satisfaction and loyalty during employee–customer interactions: The mediating role of customer affecttive commitment and perceived service quality. Cogent Bus. Manag. **5**, 8–21 (2018). https://doi.org/10.1080/23311975.2018.1491780
11. International Journal of Recent Technology and Engineering (IJRTE) ISSN: https://www.researchgate.net/publication/333446989. Accessed 20 February 2023.

A Case Study: The Transportation Challenges in Agricultural Products at Lembaga Pertubuhan Peladang (LPP), Malaysia

Mohd Hilmi Abdul Rahman, Mohd Yazid Md Taib, Hairul Rizad Md Sapry, Nik Anis Zulaikha Mohd Shaharil, and Amirul Haqim Khalid

Abstract This study is about the transportation challenges for agricultural products at Lembaga Pertubuhan Peladang (LPP). The goal of the research is to explore the challenges faced in transportation activities within the company and the effects of the lack of transportation at LPP. Moreover, interviews are conducted with the employees from LPP and PPK Kuala Langat. From the interview sessions, information was obtained about the researcher's study and proceeded to create transcripts. One of the challenges in transportation activities is the farmers produce crop yields more than demand, and transport accessibility in rural areas. The effect on the lack of transport is that it can lead to product spoilage and wastage. Based on the findings, the researcher recommends the agriculture organization implement a GPS truck-tracking system, improve road infrastructure and guide farmers to sell products using e-commerce. This research is beneficial to the LPP as it can ensure the food produced by the farmers can be sent to the market at the appointed time.

Keywords Agricultural goods · Transportation · Consolidation centre · Farmers

---

M. H. A. Rahman (✉) · M. Y. M. Taib · H. R. M. Sapry · N. A. Z. M. Shaharil · A. Haqim Khalid
Industrial Logistics, Malaysian Institute of Industrial Technology, Universiti Kuala Lumpur, Johor, Malaysia
e-mail: mhilmiar@unikl.edu.my

M. Y. M. Taib
e-mail: mohdyazid@unikl.edu.my

H. R. M. Sapry
e-mail: hairulrizad@unikl.edu.my

N. A. Z. M. Shaharil
e-mail: anis.shaharil@s.unikl.edu.my

A. Haqim Khalid
e-mail: amirulhaqim@unikl.edu.my

A. Ismail et al. (eds.), *Industrial Logistics I*,
SpringerBriefs in Applied Sciences and Technology,
https://doi.org/10.1007/978-3-032-26981-2_15

1 Introduction

As humans farm, produce, and give more time to domesticated plants or animals on a routine basis, agriculture is a kind of land use that changes the landscape [1]. The movement of individuals and items from one place to another for a variety of purposes is referred to as transportation. It performs more than just a connecting role when it comes to the distribution of finished goods and the transfer of raw materials for industry needs. One of the most crucial elements of distribution and logistics is now believed to be transportation. Although farmers are the individuals in the agriculture business who are most well-known, they are not the only ones.

The problem that LPP currently faced is the lack of necessary transportation infrastructure to handle all of the agricultural goods that the farmers have produced. The possibility of agricultural goods spoiling and being wasted is greater if delivery is delayed. It is because fresh agricultural products have to be delivered immediately. After all, it is perishable.

For distribution to the hypermarkets, LPP must handle all of the agricultural goods produced in Malaysia. The respondents stated that the farmers produce their crop yield more than the demand needed and because of that, they need more vehicles to deliver their products from the farm to the final destination. Furthermore, the business must have a functional transportation system since it may assist farmers by enabling them to sell high-quality agricultural goods [2]. Farmers and end users will be impacted if the transportation activities cannot be completed successfully.

Therefore, there are three objectives of this research:

- To explore the challenges faced in transportation activities within the company.
- To study the effects on the lack of transportation at Lembaga Pertubuhan Peladang (LPP).
- To recommend the strategy for the organization to overcome the shortage of transportation.

This study is being carried out at Lembaga Pertubuhan Peladang (LPP), which is situated in Kuala Lumpur's Jalan Sultan Salahuddin, and this research also is carried out at PPK Kuala Langat. LPP provides transportation to ease the movement of the delivery of agricultural products from farmland to the final destination. On the other hand, LPP is responsible for developing Pusat Pengumpulan Hasil Ladang (PPHP) at Pertubuhan Peladang (PP) to assist the farmers in selling their products directly to the customers. Additionally, PPHP is involved in this study as this place is where the farmers collect the crop yield. The framework for this study can be seen in Fig. 1.

The efficient planning, execution, and management of flow and related information from the place of origin to the destination of use to satisfy customer demands are known as logistics [3]. Because the agricultural sector has long been a representative of a country's economic growth, the development of agriculture has an effect on the economy [4]. As a consequence of enhanced transportation infrastructure, farmers now have a variety of options for taking part in income-generating agricultural production and market operations [5]. Although there are several limitations,

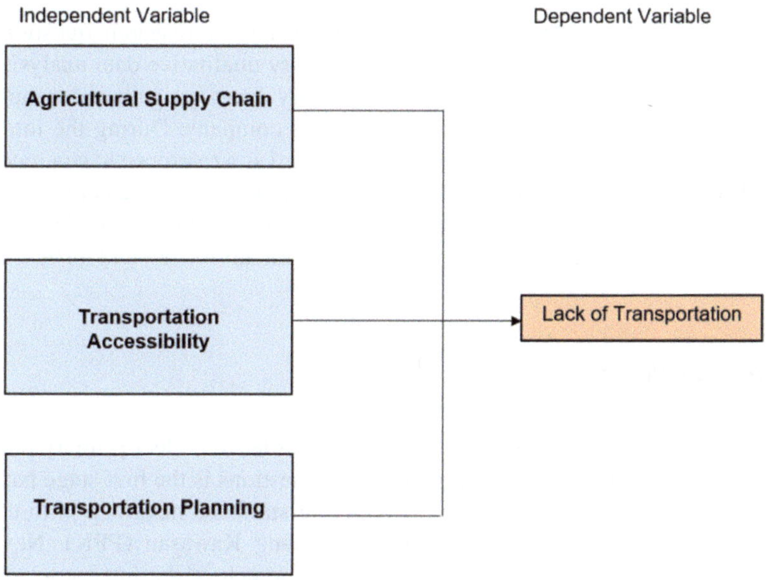

Fig. 1 Research framework

such as a lack of infrastructure and environmental protection, the connection is also a significant contributor to agricultural productivity.

Farmers must continue to produce because it is their way of life and means of support [6]. In order to improve production and profitability, roads are indeed required for all subsistence farmers in the country to have access to markets and agricultural supplies. Most people in rural areas depend on agriculture, including fisheries, crops, and livestock [7]. Then, transportation is important to transfer commodities from the farm to the markets and it is suitable for local productivity, improving local farmers' livelihood opportunities in rural areas [8].

2 Methodology

A qualitative method is used to achieve the study's objectives. The researcher could learn more about a problem or generate new study ideas with the aid of qualitative research. Both primary data and secondary data were used in this research. Primary research includes a variety of methods, including surveys, focus groups, and interviews. Observations are described as collecting detailed field information regarding what has been observed, heard, or encountered, while interviews are defined as personally asking personal questions in one-on-one communication. The questions are created by focus groups, and the conversation is among a group of individuals.

The process of using existing information to answer research questions not covered in the primary study is known as secondary qualitative data analysis such as textbooks and journals [9]. Moreover, the primary data were gathered through the interview session with three respondents from the company. During the interview session, the researcher used a voice recorder to record as a reference to generate transcripts. The people who give more information towards this research are respondent 1 who is the general manager and respondent 3 is a registered farmer under LPP. The secondary data was collected from journal and information from LPP [10].

2.1 Research Procedure

The researcher must complete five procedures in order to collect primary data and visualize relevant details. Developing interview questions is the first stage based on the literature review. In order to have a targeting staff, the next step is to contact the employees from LPP and Pertubuhan Peladang Kawasan (PPK). Next, the interviewer discusses the location, timing, and approach of the interview with the subject. The interview process included professional language and dress. Finally, the researcher can compile and evaluate the data using the T x P x L method. Figure 2 shows the research procedure.

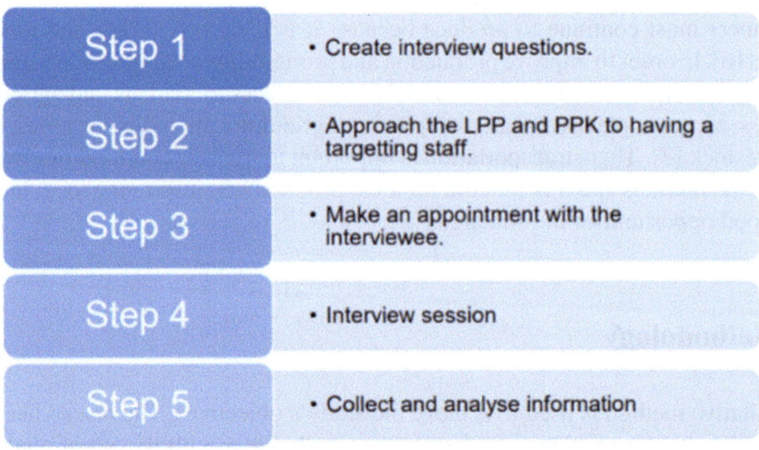

Fig. 2 Research procedure

2.2 Interview Approach

A planned discussion in which one person asks questions and the other responds is known as an interview. A one-on-one interaction between an interviewer and an interviewee is referred to as an 'interview' in normal speech. The interviewee answers the interviewer's questions by typically giving information. That info might be used to or available to additional audiences right now. All the questions are based on the literature review and the respondents answered them. Three respondents were involved in this study. They are respectively holding the position of General Manager, Head of Planning, Monitoring and Evaluation, and Head Assistance of Pertubuhan Peladang Muda.

2.3 Data Analysis

In qualitative research, data analysis refers to the methodical searching and organization of interview transcripts, observation notes, or other non-textual resources gathered by the researcher to further knowledge of the topic. When analysing qualitative data, it is necessary to read a lot of transcripts and search for similarities or variances before identifying themes and creating categories. Researchers commonly 'cut and paste' to classify the data and utilize coloured pens to find the answers. The interview is conducted through face-to-face with two respondents and one respondent through online conversation by using Microsoft Teams.

2.4 Data Analysis Method

The researcher interviewed the respondents to extract the data and information from the transcripts. To get the answers to the researcher's questions, T x P x L coding is used. Simply described, this coding aids in identifying recognized themes or subjects from the collected information for the research.

From the interview session with the respondents, the researchers will use TPL coding to interpret data and information. This format is used to ease the researcher's finding evidence on the transcript.

3 Results and Discussion

Information collected from interviews session and obtained from the targeted respondents will be analysed. All the data was collected from the organization's respondents. Additionally, there are three key steps that the researcher will engage in which

Table 1 Summary of findings

Questions	Sources
How the demand for agricultural products can affect the number of transportation needed?	The respondents stated that the demand for transportation needed is increasing when the agricultural products produced by the farmers exceed the demand. The farming product products get exceeded because the farmers plant vegetables but they do not know where to sell or market the products
In what way LPP that can help farmers to stay competitive in the agriculture sector?	To ensure the farmers stay competitive in the industry, the respondents suggested increasing more functional consolidation centres like PPK Kuala Langat and improving transportation. The guaranteed market is also crucial in this field as the farmers need to know where products will be sold
How does transportation affect farmers' output and also the product's prices?	The respondents acknowledge if the company has an efficient transportation and ensures that they can deliver the goods very quickly, it can affect the product's price. It is because when the agricultural products arrive earlier at the customers, it helps to keep the products fresh and they can demand the best price. The price of products can increase up to 10%
How are the transportation activities being operated from the point of production to the markets?	Collection centre receives agricultural products from farmers in each state. From 09:00 am to 12:00 pm, the lorries need to reach the collection centre within that time. Products are transported to factories and distribution centre such as Giant, Econsave, and Lotus. The transportation activities happen 8 times a week

are interpretation, analysis, and explanation. The research findings can be seen in Table 1.

4 Conclusion

The function of collectors and transporters is crucial for distant and smaller farms where it is not feasible for farmers to get their goods to markets using their own modes of transportation.

4.1 Reinstatement of Objectives

The researcher collects all the data and information about the lack of transportation for agricultural products at LPP. From the interview sessions, the researcher found that all respondents give different opinions.

1. To explore the challenges faced in transportation activities within the company. The challenges are transportation accessibility in rural areas and the attitude of the farmers. Sometimes, the farmers did not want to share transport.
2. To study the effects on the lack of transportation at Lembaga Pertubuhan Peladang (LPP). The researcher found that the lack of transportation can lead to agricultural products being damaged and the goods arriving late at their final destinations.
3. To recommend the strategy for the organization to overcome the shortage of transportation. The application of the GPS vehicle tracking system, the improvement of road infrastructure in rural areas and implementation of e-commerce for farmers are the strategies to overcome this issue.

4.2 Recommendation

5 GPS Truck Tracking System

The researcher suggests LPP to install GPS tracking devices on their vehicles so that they can track and provide the current location of their vehicles. Tracking cars with GPS can assist in offering real-time or historical navigation data and notifications. It will be simpler to adhere to maintenance schedules as a result. Additionally, vehicle tracking aids in avoiding any delays caused by breakdowns and will help the transporter keep its vehicle running for longer.

6 Improve Road Infrastructure

Poor road infrastructure gives difficulties for vehicles to access rural areas. The shortage of transportation infrastructure, such as roads, prevents locals from travelling to urban areas. Road infrastructure is one method for encouraging the growth of the agricultural industry in rural regions. Government needs to invest in the improvement of road infrastructure in rural areas by that vehicles can access it.

7 E-Commerce

Sometimes, the farmers produce crop yield more than the demand needs. Because of that, if the farmers cannot sell their agricultural products, they can sell them online. E-commerce is the activity of buying and selling products and services online. It is easy for the farmers to update about their goods everywhere and anytime as long as their device can connect to the internet.

Acknowledgements I am thankful to every individual who gives me support and helps me in preparing this report.

References

1. D. Harris, D. Fuller, Agriculture: definition and overview, in *Encyclopaedia of Global Archaeology*, ed. by C. Smith (Springer, New York, 2014), pp. 104–113
2. B. Gbam, Effect of transportation on the marketing of agricultural products in Jos North. J. Res. Bus. Manag. **5**, 99–106 (2017)
3. G. Gebresenbet, T. Bosona, Logistics and supply chains in agriculture and food. Swedish Univ. Agricult. Sci. Uppsala Sweden **10**, 125–146 (2012)
4. D. Xu, Research on the logistics and distribution of agricultural products in Liaoyang. Malaysian E-Commerce J. **3**, 17–20 (2019)
5. A. Ilmi, L. You, U. Sichu, R. Humphery, Agriculture production and transport infrastructure in East Africa. World Bank Group **7281**, 1–12 (2015)
6. D. Bonsu, in *Road Transport and Agriculture: A Comparative Study of the Implications of Road Access for Subsistence Agriculture in The Northern Ghana. Department of Geography* (2014), pp. 3–5
7. P. Starkey, J. Hine, *How Transport Affects Poor People with Policy Implications for Poverty Reduction. UN-Habitat, the Overseas Development Institute* (2014), pp. 4–52
8. I. Abdulraheem, T. Adefare, Impact of transportation on agricultural practices and production in rural areas. Implicat. Sustain. Food Secur. **35**, 27475–27479 (2021)
9. T. Sutehall, M. Sque, J. Hall, Secondary analysis of quantitative data: a valuable method for exploring sensitive issues with an elusive population. J. Res. Nurs. **16**(4), 335–344 (2011)
10. L.P. Peladang, *2016 Annual Report* (2016)

Understanding the Islamic Hotel Concept from the Perspectives of Hotel Operators in Malaysia

Nor Amira Mohd Razali, Harnidah Samengon, Nur Hafizah Muhammad, and Nurashikin A. Ridzuan

Abstract The increasing global Muslim population has resulted in an increase in Muslim travelers worldwide, hence increasing the demand for Muslim-friendly products and services in the hospitality industry. However, limited research has examined how hotel operators, as the key implementers, understand and interpret the Islamic hotel concept. A qualitative approach using semi-structured interviews with five hotel operators across the Peninsular Malaysia was employed to explore the hotel operators' comprehension and operationalization of Islamic hotel principles. The thematic analysis revealed five key themes namely full implementation of Islamic practices; provision of halal food and Muslim-friendly facilities; Shariah-compliant operations and financing; adherence to Allah's and the Prophet's (PBUH) teachings; and gaining Muslim guests' trust and confidence. This study contributes to the body of knowledge on Islamic hospitality management and underscores the importance of enhancing operators' understanding to ensure genuine adherence to *Shariah*-compliant practices in Malaysia's hotel industry.

Keywords Islamic hotel · Muslim-friendly tourism · Hotel operators · Islamic tourism · Perception

N. A. M. Razali (✉) · H. Samengon · N. H. Muhammad · N. A. Ridzuan
Faculty of Hospitality, Tourism and Wellness, Universiti Malaysia Kelantan, Kota Bharu, Malaysia
e-mail: noramira.mr@umk.edu.my

H. Samengon
e-mail: harnidah@umk.edu.my

N. H. Muhammad
e-mail: hafizahm@umk.edu.my

N. A. Ridzuan
e-mail: nurashikin.ar@umk.edu.my

A. Ismail et al. (eds.), *Industrial Logistics I*,
SpringerBriefs in Applied Sciences and Technology,
https://doi.org/10.1007/978-3-032-26981-2_16

1 Introduction

Islamic tourism is one of the promising market segments in the global tourism industry that cater to both Muslim and non-Muslim markets [1]. It has emerged as one of the fastest-growing sectors within tourism and hospitality [2]. The drivers behind this market growth include the growing Muslim population, an increasingly youthful demographic, rising disposable incomes among the middle class, the availability of travel information, and specific travel types such as Ramadan and business travel [3].

Globally, Muslim tourists represent the second-largest population with approximately 2.12 billion individuals [4]. Muslims accounted for 110 million global tourist arrival in 2022 which represent approximately 68% of the pre-pandemic levels from 2019. The global Muslim tourist arrivals are projected to increase to 230 million by 2028 with estimated expenditure of USD 225 billion [5]. Despite the potential it has, Islamic tourism remains in its infancy and requires further exploration to develop its foundational knowledge [6].

Muslims constitute a significant market for various industries including pharmaceuticals, logistics, finance and hospitality [7]. Unlike other tourists, Muslim tourists have unique needs that governed by their religious obligations [8]. These needs include access to halal food, prayer facilities, gambling and alcohol-free environments, and adherence to Islamic moral principles [9]. Those business operators that able to meeting these needs have better chances to attract this lucrative market.

In Malaysia, the country welcomed 10.07 million international tourists, including 2.12 million Muslims who contributed RM5.37 billion in expenditures in 2022 [10]. Malaysia's unique multiracial identity and Islam as its official religion enhance its appeal as a destination for Muslim travelers. Notably, Malaysia has maintained its top ranking in the Global Muslim Travel Index since its launch in 2015 [5].

Islamic tourism fosters interfaith understanding and strengthens relationships across different races [11, 12]. Islamic hospitality, as part of Islamic tourism, provides travelers with the opportunity to experience the modesty and uniqueness of the Muslim lifestyle. Accommodation, such as hotels, motels, and resorts, plays a crucial role in meeting the lodging needs of Muslim tourists after daily touring activities. To cater to this niche market, Malaysia has prioritized the development of Muslim-friendly facilities and services.

The Malaysian government has aggressively promoted Muslim-friendly tourism and hospitality to attract Muslim tourists [13]. Initiatives include the launch of the Muslim-Friendly Accommodation Recognition (MFAR) program in 2018, which was rebranded as the Muslim-Friendly Tourism and Hospitality Assurance and Recognition (MFAR) in 2023 under the Islamic Tourism Center (ITC). Hotels recognized under the MFAR program are classified into Silver, Gold, or Platinum tiers based on their compliance with Islamic practices.

Most existing studies on Islamic hotels primarily focus on awareness and acceptance of Islamic hotel concepts among operators and guests [14], implementation of Islamic hotel concepts [15], examining barriers and challenges in implementation [16, 17], and frameworks for certification [12, 18]. However, there is a lack of research that emphasizes the perspectives of hotel operators, the key implementers of this concept, in understanding and translating Islamic hotel principles into practice. Therefore, this study aims to investigate the understanding of Islamic hotel operators regarding the Islamic hotel concept in Malaysia. By addressing this objective, the study provides valuable insights into how the concept is interpreted and operationalized within the industry, contributing to the development of more effective guidelines and practices that align with the true essence of Islamic hospitality.

2 Methodology

A qualitative research approach using semi-structured interviews was utilized to explore the understanding of the hotel operators on the Islamic hotel concept. The interview sessions were guided by interview protocol throughout the data collection process to ensure the focus, consistency, and reliability of the study [19].

This study adopted the purposive sampling strategy which involved five hotel operators that represent Islamic hotels across the Peninsular Malaysia. These interviewees were selected based on a few criteria like held decision-making authority and actively involved in hotel operations, industry experience and tenure in the hospitality sector to ensure they have informed perspectives on the Islamic hotel concept and its implementation. The demographic profile of the interviewees is shown in Table 1.

All interviews were conducted face-to-face and audio-recorded with interviewees' consent. Prior to each session, interviewees were informed of the study's purpose and assured of confidentiality to promote openness and trust. The recordings were

Table 1 Demographic profile of the interviewees

Interviewee	Gender	Position	Organization (location)	Islamic practices implementation period (Year)
Interviewee 1	Male	General Manager	Islamic Hotel A1 (Selangor)	11
Interviewee 5	Female	Director of Sales Marketing	Islamic Hotel A2 (Selangor)	4
Interviewee 2	Male	General Manager	Islamic Hotel B1 (Terengganu)	8
Interviewee 3	Male	Quality and *Shariah* Compliance Executive	Islamic Hotel B2 (Kelantan)	7
Interviewee 4	Male	Islamic Division Executive	Islamic Hotel C (Kedah)	7

transcribed verbatim and analyzed using thematic analysis, following the six-phase framework [20]: (a) data familiarization, (b) generating initial codes, (c) searching for themes, (d) reviewing themes, (e) defining and naming themes, and (f) producing the report.

This systematic process facilitated a comprehensive understanding of the participants' views, offering valuable insights into how Islamic hotel operators conceptualize and implement Islamic hospitality practices in Malaysia.

3 Results and Discussion

The findings reveal a variety of perspectives among hotel operators regarding their understanding of the Islamic hotel concept. Overall, the participants' views were closely aligned with the principles prescribed in the *Quran* and *Hadith*, reflecting their efforts to incorporate Islamic values in hotel operations. When asked about their understanding of the Islamic hotel concept, five major themes emerged from the interviews: (1) full implementation of Islamic practices, (2) provision of halal food and Muslim-friendly facilities, (3) Shariah-compliant operations and financing, (4) adherence to Allah's and the Prophet's (PBUH) teachings, and (5) building Muslim guests' trust and confidence.

The first theme which is full implementation of Islamic practices in hotel properties is exemplifies by the General Manager of a hotel located in Shah Alam, Selangor as follow:

> ...Islamic hotel, we use Muslim friendly hotel term. ITC has given few categories, depending on the hotel practices on the Muslim-friendly attributes. At our hotel, we practice, insha Allah, 100% or 99% where we get platinum category award. Among the attributes are restaurants that do not serve liquor, serve halal food, uniform of the staff for Muslim staff that meets the shariah compliance. In terms of rooms, it makes it easier for people to take ablution, availability of Qibla direction, Quran available upon request. But now Qibla can use apps, right, so it cannot be the criteria anymore. Facilities like swimming pool, there is time separation for use. For ladies only, the staff takes care of the ladies. Our swimming pool is covered, there is a door, so that is an advantage for us besides separation. In addition, guests need to wear a bathing suit to cover their aurah. Our spa is for ladies only. Loans for finance must be shariah compliance. But so far, I have seen, ITC does not have that requirement in the accreditation
>
> [Interviewee 1]

In the excerpt, he believed that he had successfully adhered to 99% to 100% of the Muslim-Friendly Tourism and Hospitality Assurance and Recognition (MFAR) guideline established by the Islamic Tourism Center. This guideline outlines a number of Islamic practices that hotel operators must follow in order to be recognized as an MFAR hotel. As such, the General Manager believed that he had a comprehensive understanding of the Islamic hotel concept based on his practical implementation of the MFAR guidelines.

The second theme that emerged from the interviews pertained to the provision of *halal* food and facilities that cater to Muslim needs, an essential criteria indicating the

Islamic hotel concept. The General Manager of a hotel located in Kuala Terengganu, Terengganu said that:

> …basically, Islamic hotel symbolizes of halal, and the facilities offered need to complement the Muslim's perception as well as their acceptance.
>
> [Interviewee 2]

He asserted that an Islamic hotel was a symbol of *halal*, which required providing facilities that could accommodate the needs and preferences of Muslim guests.

The third theme is *halal* symbolization and facilities offerings that fulfill the Muslim needs with addition to *shariah* compliance financing, operation, and activities. This theme illustrates that the concept of Islamic hotels has widen to include *halal* symbolization and facility offerings that cater to the needs of Muslim guests. In addition, some hotels follow *Shariah* law in its financing, operations, and activities. As explained by the Quality and *Shariah* Compliance Executive of a hotel located in Kota Bharu, Kelantan as follow:

> The usual hotel categories are conventional, Muslim-friendly hotels and shariah compliance hotels…Muslim-friendly hotel is just introduced under MOTAC in 2016. It focuses on Muslim customers. The majority of hotels in Malaysia are Muslim-friendly because one of their criteria is the provision of halal to guests. It is not necessary that all outlets must be halal. As well as some other criteria that have been set by the ITC…for an Islamic or shariah-compliant hotel I would say, it has little difference compared to other hotels, especially from a financial point of view as well as other operations and activities
>
> [Interviewee 3]

The excerpt exemplifies that Islamic hotels encompass not only *halal* food but all aspects of hotel operations, including financial management, operational procedures, and leisure activities. I guess the idea behind the Islamic hotel concept is to provide an environment that is in line with Islamic values and principles and caters to the needs of Muslim travelers.

The fourth theme underpins Allah and the Prophet (PBUH) as the guideline in defining the Islamic hotel concept. This theme highlights the overarching commands of Allah and the Prophet's (PBUH) wishes in defining the Islamic hotel concept. As stated by the Islamic Division Executive of an Islamic hotel located in Langkawi, Kedah as follow:

> … a hotel that practices Islamic teachings as being told by Allah or Prophet Muhammad (PBUH)…Islamic hotel can be divided into two (categories), either shariah compliance hotel or Muslim-friendly hotel. Both hotels can be considered as Islamic hotel, it is just that shariah compliance hotel practicing Islamic value more strictly than Muslim-friendly hotel.
>
> [Interviewee 4]

In the excerpt, he stated that an Islamic hotel was the one that followed the teachings of Islam as prescribed by Allah and the Prophet (PBUH) in its operations.

Lastly, the final theme focuses on gaining Muslim guests' confidence and trust. The Director of Sales and Marketing of an Islamic hotel located in Sepang, Selangor emphasized the importance of building customer trust and confidence as follow:

> … a hotel is called Islamic hotel when a Muslim guest doesn't feel worry to get into the hotel. Didn't feel suspicious whether the food contained alcohol or not. Like steak, the original recipe does use wine, but for Islamic hotel, you know for fact that the ingredient won't be there…certain hotel that you go, there is no qiblah sign. But, when you go to either Muslim-friendly or Islamic hotel, everything is there, qibla sign, telekung, Quran.
>
> [Interviewee 5]

According to her understanding, the Islamic hotel concept was achieved when Muslim guests had confidence in staying and dining at her hotel. This confidence came from the assurance that the food served is *halal* and that the facilities and services provided fulfill their religious obligations. Thus, it can be summarized that the hotel operators' understanding is consistent with what is preached by *Quran* and narrated by *hadith* as found in the first objective.

Overall, these themes demonstrate that hotel operators' understanding of the Islamic hotel concept is largely consistent with Islamic teachings. Participants recognized the importance of aligning operations with *Shariah* principles and providing an environment conducive to Muslim guests' comfort and confidence.

The findings also indicate that two main categories of Islamic hotels exist in practice which are *Shariah*-compliant hotels and Muslim-friendly hotels. The distinction lies in the extent of *Shariah* application, although both terms are often used interchangeably. This suggests a need for standardization and clearer categorization to avoid confusion among industry practitioners.

In contrast to the findings [14], who reported limited understanding among hotel operators regarding the Islamic hotel concept, the present study found that participants demonstrated substantial awareness and comprehension of Islamic hospitality practices. These insights are consistent with previous definitions which described Islamic accommodations as those offering Shariah-compliant services beyond halal food and beverages [17], and by [21], who associated "halal hotels" with facilities such as separate prayer areas and gender-segregated swimming pools. Together, these findings confirm that the hotel operators in this study possess a strong understanding of Islamic hotel principles and their operationalization in the Malaysian context.

4 Conclusion

As a conclusion, it is crucial to recognize the level of knowledge that hotel operators possess as it significantly shapes their actions and decision-making. It is emphasized that hotel operators whom have strong understanding about Islamic hotel concept are more likely to implement the Islamic practices within their establishments [14]. Conversely, a lack of awareness or understanding of the importance of adhering to *Shariah* principles often contributes to the limited application of Islamic values in the hospitality industry. This highlights the need to enhance awareness and comprehension of Islamic hospitality principles among hotel operators, ensuring that Islamic hotel practices are implemented in alignment with the true essence of *Shariah*-compliant hospitality.

Acknowledgements This work is funded by the Universiti Malaysia Kelantan, under the UMK-FUND research grant (R/FUND/A1100/01519A/001/2024/01236). The authors would like to express gratitude to the organization above and to all who directly or indirectly contributed to the successful of this publication.

References

1. J. Juliana, W.A.J. Mahri, A.R. Salsabilla, M. Muhammad, I.S. Nusannas, The determinants of Muslim millennials' visiting intention towards halal tourist attraction. J. Islamic Account. Bus. Res. **14**(3), 473–488 (2023). https://doi.org/10.1108/JIABR-02-2021-0044
2. A. Mursid, P. Anoraga, Halal destination attributes and revisits intention: The role of destination attractiveness and perceived value. Int. J. Tour. Cities **8**(2), 513–528 (2022). https://doi.org/10.1108/IJTC-03-2021-0040
3. Mastercard-Crescentrating Global Muslim Travel Index 2022. Crescent Rating Pte. Ltd. https://www.crescentrating.com/reports/global-muslim-travel-index-2022.html (2022). Accessed 10 October 2025
4. Mastercard-Crescentrating Global Muslim Travel Index 2024. Crescent Rating Pte. Ltd. https://www.crescentrating.com/reports/global-muslim-travel-index-2024.html4 (2024). Accessed 10 October 2025
5. Mastercard-Crescentrating Global Muslim Travel Index 2023. Crescent Rating Pte. Ltd https://www.crescentrating.com/reports/global-muslim-travel-index-2023.html (2023). Accessed 10 October 2025
6. A. Vargas-Sanchez, M. Moral-Moral, Halal tourism: literature review and experts' view. J. Islamic Market. **11**(3), 549–569 (2019). https://doi.org/10.1108/JIMA-04-2017-0039
7. A. Ali, G. Xiaoling, M. Sherwani, A. Ali, Antecedents of consumers' halal brand purchase intention: an integrated approach. Manag. Decis. **56**(4), 715–735 (2018). https://doi.org/10.1108/MD-11-2016-078
8. O.K. Bhatti, U.S. Aslam, A. Hassan, M. Sulaiman, Employee motivation: an Islamic perspective. Humanomics **32**(1), 33–47 (2016). https://doi.org/10.1108/H-10-2015-0066
9. M. Battour, M.N. Ismail, M. Battor, M. Awais, Islamic tourism: an empirical examination of travel motivation and satisfaction in Malaysia. Curr. Issue Tour. **20**(1), 50–67 (2017)
10. Rest and Rehlah. Islamic Tourism Centre. https://itc.gov.my/rest-rehlah-sept-oct-2023 (2023). Accessed 6 September 2023
11. K.H. Din, Islam and tourism: patterns, issues and options. Ann. Tour. Res. **16**(4), 542–563 (1989). https://doi.org/10.1016/0160-7383(89)90008-X
12. M.R. Razalli, R.Z. Yusoff, M.W. Roslan, A framework of *Halal* certification practices for hotel industry. Asian Soc. Sci. **9**(11), 316–326 (2013). https://doi.org/10.5539/ass.v9n11p316
13. A.S. Rusydiana, F.F. Hasib, S.S. Rahayu, Islamic hotel indicators: a bibliometric study. Int. J. Relig. Tour. Pilgrimage **9**(1), 13 (2021). https://doi.org/10.21427/8aa4-qw88
14. N.A. Zainol, R. Ahmad, M.S.K. Ideris, N.R.A.N. Azam, Assessing the awareness and perception of hotel operators towards *shariah* compliant hotel operations. Int. J. Serv. Manag. Sustain. **4**(1), 85–102 (2020). https://doi.org/10.24191/ijsms.v4i1.8061
15. M.H.A. Karim, R. Ahmad, N.A. Zainol, Differences in hotel attributes: Islamic hotel and Sharia compliant hotel in Malaysia. J. Glob. Bus. Soc. Entrepreneurship (GBSE) **1**(2), 157–169 (2017)
16. H.E. Saad, B.N. Ali, A.M. Abdel-Ati, Sharia-compliant hotels in Egypt: concepts and challenges. Adv. Hospital. Tour. Res. **2**(1), 1–15 (2014). https://dergipark.org.tr/en/pub/ahtr/issue/32308/359044
17. Z. Samori, F. Abd Rahman, Towards the formation of *Shariah* compliant hotel in Malaysia: an exploratory study on its opportunities and challenges, in *Proceedings of WEI International Academic Conference 2013, Istanbul, Turkey* (2013)

18. M. Park, M.A. Jamaludin, A framework of *Halal* industry support system in non-muslim country: focusing on South Korea. J. Halal Industr. Serv. **1**(1), (2018). http://journals.hh-publisher.com/index.php/JHIS/article/view/7/30
19. R.K. Yin, *Applications of Case Study Research* (Sage, Thousand Oaks, 2011)
20. V. Clarke, V. Braun, Thematic analysis. J. Positive Psychol. **12**(3), 297–298 (2017). https://doi.org/10.1080/17439760.2016.1262613
21. A. Mohsin, N. Ramli, B.A. Alkhulayfi, Halal tourism: emerging opportunities. Tour. Manag. Perspect. **19**, 137–143 (2016). https://doi.org/10.1016/j.tmp.2015.12.010

MIX
Papier aus verantwortungsvollen Quellen
Paper from responsible sources
FSC® C105338

If you have any concerns about our products,
you can contact us on
ProductSafety@springernature.com

In case Publisher is established outside the EU,
the EU authorized representative is:
Springer Nature Customer Service Center GmbH
Europaplatz 3, 69115 Heidelberg, Germany

Printed by Libri Plureos GmbH
in Hamburg, Germany